Dialogues on
Modern Physics

Dialogues on Modern Physics

Mendel Sachs

SUNY, Buffalo

World Scientific
Singapore • New Jersey • London • Hong Kong

Published by

World Scientific Publishing Co. Pte. Ltd.

P O Box 128, Farrer Road, Singapore 912805

USA office: Suite 1B, 1060 Main Street, River Edge, NJ 07661

UK office: 57 Shelton Street, Covent Garden, London WC2H 9HE

Library of Congress Cataloging-in-Publication Data
Sachs, Mendel.
    Dialogues on modern physics / Mendel Sachs.
      p.   cm.
    Includes bibliographical references and index.
    ISBN 9810231911
    1. Quantum theory.   2. Relativity (Physics)   I. Title.
QC174.12.S224   1998
530.1--dc21   98-5083
                                                 CIP

British Library Cataloguing-in-Publication Data
A catalogue record for this book is available from the British Library.

This book is printed on acid-free paper.

Printed in Singapore by Uto-Print

To the memory of my favorite brother

DAVID

Thank you for teaching me how to see humor in the face of crisis.

It is our responsibility as scientists, knowing the great progress which comes from a satisfactory philosophy of ignorance, the great progress which is the fruit of freedom of thought, to proclaim the value of this freedom; to teach how doubt is not to be feared but welcomed and discussed; and to demand this freedom as our duty to all coming generations.

> R. P. Feynman
> from: **What do You Care What Other People Think?**
> (Norton, 1988), p. 248.

For when something has been demonstrated, the correctness of the matter is not increased, and certainty regarding it is not strengthened by the consensus of all men of knowledge with regard to it. Nor could its correctness be diminished and certainty regarding it be weakened even if all people on Earth disagreed with it.

> Moses Maimonides
> from: **Guide of the Perplexed**
> transl. S. Pines (University of Chicago Press, 1963), p. 290.

Contents

Acknowledgements

This book evolved from an Honors course that I have been teaching at my home university, State University of New York at Buffalo, for the past several years. It is a course designed for all undergraduate majors, to enhance their understanding of concepts in modern physics as well as helping to increase their powers to think creatively.

The final form of the manuscript was completed when I was on sabbatical leave at the University of California, Los Angeles, in the 1993–94 academic year. As a part of this course, on chosen days of the term, students acted out some of the dialogues while the remainder of the class was invited to interact with the 'actors.' I was convinced that this was a very effective way to teach (sometimes asserted) difficult concepts of modern physics; indeed, I saw in this experience, as a teacher, the wisdom of Plato's and Galileo's dialectical approach to exploring scientific truth.

I wish to thank my students at UCLA who participated in the Honors course, HC 33: 'From Aristotle to Einstein,' in the UCLA spring quarter, 1994, as well as my SUNY/Buffalo students, for their previous participations. Their excellent and uninhibited questions, largely promoted by the dialogue format, enhanced the flow of ideas in the discussions with much vigor and new curiosity, and, most of all, I believe they learned the positive value of 'controversy' toward the search for scientific truth.

Special thanks are due to my colleague from the Department of English, Professor James Bunn, for his most insightful comments on a reading of the first draft of the manuscript. I am also indebted to Ms. Carolyn Haist for her excellent preparation of the final form of the typescript.

Mendel Sachs
Department of Physics
State University of New York at Buffalo

Preface

A unique occurrence in the history of science that took place in the twentieth century was the simultaneous discoveries of two major upheavals in scientific thought — the quantum and relativity theories. For in the past, only one scientific revolution happened at a time.

Most exciting about the simultaneous appearances of the quantum and relativity theories in our time is that on the one hand, to establish their truth values in science, each of these theories requires a total unification with the other for its completion, on its own terms! But on the other hand, these theories have never been unified in a totally satisfactory manner, either mathematically or logically, because of some fundamental incompatibilities between them.

This is the major dichotomy of contemporary physics that must be resolved before real progress can occur — in spite of all of the activity that is taking place in this field. The dichotomy implies that either a) one of these theories must be abandoned for the other, which must then be generalized in such a way as to incorporate the empirically correct features of the abandoned approach, or b) both theories should be abandoned and replaced with an entirely new theory, that could successfully duplicate all of the correct features of both the quantum and relativity theories, and then go further. The history of science reveals that the most probable course would be the former one, utilizing the so-called principle of correspondence, applied to the way in which the accepted theory must approach the form of the abandoned theory, where it had been empirically successful.

Whichever will be the future course of physics, it is important for all of us to be aware of the conceptual differences between these two major developments in contemporary physics. It is, of course, also essential for the professional scientists to understand, in addition, the mathematical differences between these two theories. The aim of this book, however, is only to discuss the conceptual differences, hopefully in a maximally understandable and non-mathematical way.

The differences between the quantum and relativity theories, as competing explanations for the behavior of matter, is in regard to two of their essential aspects. First, they are based on contrasting physical models — implying different ontologies. Second, they are based on contrasting epistemological outlooks — that is, in dealing with the limit of knowledge that we may achieve.

The method of presentation of these contrasting ideas will be in the dialectical style of dialogues. This method was most effective in the dialogues of Plato, in ancient Greece, on questions about mathematics, physics and philosophy, education, government and epistemology. More than any other author, in my view, Plato helped to initiate an evolution of the ideas of the Western Civilization, to the present time.[1]

Another set of efficacious dialogues in the history of science were those of Galileo, in the Renaissance period, 19 centuries after Plato. His main works, entitled, *Dialogues Concerning Two Chief World Systems* and *Two New Sciences*, were extremely effective in launching ideas and a methodology in physics research that indeed have persisted to the present time. It is for this reason that Galileo is sometimes referred to as 'the father of modern physics.'[2]

In our own time, a very useful dialogue was the one written by the theoretical physicist J. M. Jauch, entitled *Are Quanta Real?*, dealing with the philosophy of the quantum theory.[3]

As in these previous works, I have chosen for the dialogues three characters, whom I have named: Manny, Mo and Jackie. They engage in dialogues on some of the main ideas of the quantum and relativity theories, and their applications to some of the problems in elementary particle physics of the smallest domain of physics experimentation, and to its largest domain — the physics of the universe as a whole, cosmology, that is 42 powers of ten greater than the domain of particle physics, with respect to our present day experimentation.

Manny, Mo and Jackie are the three theoretical physicists of a ten-member Physics Department of a small liberal arts college in western Oregon, Leber College. They meet once a week, for the full eight-week Fall quarter of the academic year. Manny takes one point of view about the ideas of physics and its related philosophy (which is the consensus view today, in this last decade of the 20th century), Jackie takes views opposite to those of Manny, and Mo mediates their argument, sometimes siding with Manny and sometimes with Jackie. It is important to note that Manny, Mo and Jackie come to their ideas strictly through intellectual adventure, never to only please others in conforming with one group or another.

It is not my aim to convince the reader of one side of these arguments or the other. The main object is to present the conflicting views as clearly as possible on some of the essential ideas of contemporary physics and the philosophy of science, and then to encourage the readers to decide for themselves about the most plausible path toward scientific truth.

Manny, Mo and Jackie are very close friends who have become aware of the value of intellectual debate and their agreement to disagree, and to respect each others opinions, even when strongly disagreeing with them. It is my hope that the sense of excitement and adventure in the pursuit of ideas about the universe, in any of its domains, that is experienced by these three colleagues in their dialogues, will carry over to the reader.

FOREWORD

Georges Lochak
Fondation Louis de Broglie
Paris, France

One may consider, as does the author of this book, that dialogue is a means of reaching the truth. But one may also consider dialogue simply as a literary genre, a means of realizing the truth (the truth of the author) in a more subtle and less tedious way than a dogmatic treatise; in pretending to give an adversary his say by permitting him to present his point of view which one can then confront. If I say 'pretend' it is not because I suppose that the author of a dialogue tries to trick us, but because in the final analysis one should not forget that the author knows what he wants; and what he wants is to convince the reader. This is why, in all dialogues, one always recognizes, among the protagonists, the one who holds the position of the author and the ones who represent the opposing camp.

Thus, in the *Dialogues Concerning Two Chief World Systems*, by Galileo, one knows at the outset that Galileo is Salviati, the Copernican, Simplicio, is the 'other,' i.e. the Aristotelian. As for Sagredo, the wise gentleman who plays the role of arbitrator, he is not really wise for seeing the light since no one doubts that this light comes from Salviati. That Galileo had trouble with the censors was not due to the ideas of Simplicio, but those of Salviati.

In the present case it is the same in that a few pages of reading are enough to know that Sagredo, the wise gentleman, is Mo; the Aristotelian, or should I say the proponent of the quantum theory, is Manny; and that Salviati is Jackie. The dialogues that the reader has before his eyes are good because they follow the law of the genre and this is essential. They are good because they review with a critical eye all of contemporary physics. And especially, when one has just begun to read them, one feels a desire to debate with all three protagonists. The proof of this is that I began my Foreword contesting the author's ideas. But he should be reassured that I am one of his friends. I know that he is a realist, a determinist, an advocate of continuity rather than discreteness, that he thinks, as Galileo, that the book of Nature is written in a mathematical language, and that the world obeys definite laws and that chaos is only apparent, due to our momentary ignorance. I will admit, nevertheless, that while usually being on Jackie's side, I do not share all her views. I surely share even less with Manny, given his position of empiricism, with a disdain for the internal logic of theories and his belief in fundamental indeterminism. Like Jackie, I believe in the existence of an objective reality, admitting fully, as she does, that this is only a belief. But at the same time I am not certain whether the laws of physics are laws of Nature and that these laws should be expressed in the language of mathematics. I rather have the impression that this is our book, for us, that we have written in this language, and that it is we who project the physical laws onto the mystery of Nature. The proof is that we are frequently obliged to change the laws, as well as the fact that several different laws may account for the same phenomenon, which shows that each of them renders only a small facet of reality; no law is true to all reality. At least this is the way I see things. In effect we invent laws to give sense to the 'shadows in the depths of Plato's cave.' But even though these are our laws and not those of Nature, and they are not, as I feel, in effect outside of the cave, I believe profoundly that if our laws sometimes give good results, then they have an ontological quality. There must be within them some objective truth since they permit us to interact with Nature, which in turn seems therefore to understand our language.

As for the relation of physics and mathematics discussed in the book, I hold Einstein's view that "as much as mathematical propositions relate to reality, they are not certain, and as much as they are certain, they do not relate to reality." As for mathematics itself, I wish here also to engage briefly in this discussion to say that, like Jackie, I believe that mathematics is not science, though only in the sense of the science of Nature. For me, mathematics is a science concerned with the objects of the spirit. But from a certain point of view, even the natural sciences deal with objects created by the spirit; in this case they are the concepts by which we represent natural objects. The difference between the two is that in the natural sciences one tries at the outset to connect a concept with the observations; whereas in mathematics, one is given the liberty to explore the mathematical domain with no restraints except for logic. As for the selection of a mathematical object, this is frequently guided by physics. And even if it isn't, it is not arbitrary and proceeds in general by means of a new way of viewing existent objects. In other words, known mathematical objects offer to the investigation a field of study and expansion completely analogous to that offered in physics by a range of natural objects.

The author and the reader should pardon me because instead of confining myself to the role of Foreword writer, I have engaged in dialogue with the book (in having the advantage over the reader in having read it). But I am sure that when he reads it, the reader will do the same and won't stay neutral in the face of contradictory points of view expressed in incisive fashion, concerning all of the problems of modern physics. How can one not appreciate questions such as Jackie's on the subject of light, when she contests its interpretation as the potentiality of photons, when she asks, "But what is it that propagates at 300,000 km/sec?" And when one reads the remarkable passage about relativity and the twin paradox, how can I refrain from adding my two cents worth in citing an anecdote recounted to me by the great relativist, Marie-Antoinette Tonnelat? At a meeting she circulated a questionnaire, to which the participants of the meeting were asked to respond under the cover of anonymity. The question was: Is the twin paradox taken up in special relativity or general relativity? There were three possible responses:

1) In special relativity
2) In general relativity
3) I don't know

The proportional breakdown of the responses was one third for each of these possible answers. I would like to point out that Madame Tonnelat as well as myself picked answer #2, though some eminent friends of mine picked

#1, and many others picked #3. (In any case, I think that it was supposed to turn out this way).

I would also very much have liked to participate in the discussions of holism, quantum field theory, elementary particles, entropy and the unitary field. But I cannot write so extensively in the framework of a Foreword. I only wish to make two or three additional remarks, some of which concern Louis de Broglie.

The first point deals with the principle of complementarity. Like Jackie, I detest the idea that an atom can be wave or particle (fish or fowl), according to the way that one sees it. But I would like to remind the reader that Louis de Broglie had a completely different view, one in which a singular wave, by virtue of its singularity, represents the particle aspect of the quantum object. In this way, the particle becomes intrinsically bound to the wave and guided by it. In the interference or diffraction, it is the wave that will undergo the effects according to limiting conditions, and its distortion will modify the trajectory of the singularity — the phenomenon that is revealed through observation of this singularity that carries away almost all of the energy of the singular wave and alone is capable of leaving an impression on photographic plates or in counters. In this view, the matter is therefore at once wave and particle and one should note that this was proposed by de Broglie in 1925, two years before a similar view was proposed in general relativity by Einstein and Grommer.[1]

My second remark concerns nonlinearity. When de Broglie reviewed his ideas on wave-particle dualism, in 1950, he understood it to demonstrate a theorem analogous to that of Einstein and Grommer on piloting singularities and, above all, to represent the atom by an 'undispersed wave' (a 'soliton'), in which he needed nonlinear wave equations in wave mechanics.[2] The de Broglie group was one of the first, and for many years, the only one in the world to do research on nonlinear wave equations. Unfortunately this research probably still lacks a new fundamental principle and has not been successful to this time.

As for the idea of such research, that runs counter to the dominant directions, I would like to end, like Jackie, with the idea that the value of an idea in science does not depend on the number of people who uphold it. On this subject there is a beautiful anecdote from Einstein. In his youth, when relativity was still very violently opposed, a declaration appeared, signed by 120 physicists, claiming that relativity was false. Einstein remarked, "If relativity were false, the signature of one professor would have been sufficient."

WEEK 1

ON QUANTUM RANDOMNESS

The campus at Leber College in Oregon was quiet at four o'clock in the afternoon. It was a beautiful fall day, without any hint of the deluge of rain that just ended a day ago. The students were mostly in their dormitories at this hour, or else they were working in the central library. The crisp sounds of the chirping of the robins and sparrows had replaced the calls and laughs of the undergraduates throughout this small campus.

The Conference Room of the Department of Physics had two spectacular views, one of them was westward, overlooking the campus greenery and trees, the quaint red brick buildings and the ever-rushing Willamette River along the western edge of the campus grounds. Through the window on the east side of the Conference Room could be seen the snow capped mountains of the Cascade range, etched out against a clear azure sky.

Unwinding from a lecture on intermediate modern physics, Dr. Manual Cooper still had a trace of chalk dust on the sleeves of his tweed jacket, as he stared at the lovely westward view of lush green grass and mammoth trees, buildings and water. His concentration was interrupted when his colleague, Dr. Jackie Smith, entered the large empty room, joining into the focus on Oregon countryside.

JACKIE: Well, Manny, do you see what I meant last week when I talked about the beauty in the continuity of Nature? Doesn't this view prove to you that the world is basically continuous, rather than atomistically discrete?

MANNY: What I see, Jackie, is a set of discrete things, not continuity! There is a tree over there, and next to it on the ground a bunch of chestnuts that fell from it. This is an atomistic world, if I ever saw one!

Not aware that their colleague, Dr. Maurice Linder had entered the room while they were talking, and was focusing his attention on the mountains in the eastward direction, he answered them.

MO: Well, my friends, maybe you are both correct! — poetically speaking that is. I mean, if you want to look for continuity in the world, Jackie, this is what you would see as an underlying reality, and if you, Manny, look for the atomistic aspect of the real world, this is what would be real for you. It is like an Escher painting — one sees white geese flying eastward, but a change of attitude of the observer would suddenly make the view change to black geese flying in the westward direction! Look, Jackie and Manny, you're both right — it is like wave-particle dualism in modern physics.[1]

Manny, Mo and Jackie were physicists and colleagues in the Physics Department of Leber College. They had been meeting weekly at four o'clock on Tuesdays in the Conference Room of the Physics Department. They found it refreshing to engage in intellectual meanderings that took them far from the humdrum routines of everyday living. These meetings allowed the three of them to have some exciting adventures into the world of ideas in science, beyond the commonly accepted ideas that filled their books. They decided that no holds should be barred and they would throw all possible ideas onto the table, old and new, chew them and either digest them or spit them out, without worrying about what the leaders in the field and their followers would say if there was any heresy in the meanderings. Neither did they get angry at each other, in spite of the heated disagreements at times.

On this day it seemed that they were going to debate the reality of the quantum world according to modern physics. The mountains and the rivers, the trees and the fields of greenery, suggested it in the strangest way — for Jackie saw Nature in the way of continuity and Manny saw Nature in the way of particularity, seemingly quite opposite from Jackie's view, while Mo had the schizophrenic view of reality that accepted the truth of Jackie's view and Manny's view, simultaneously. As he was pouring steaming black coffee for the three of them, Manny remarked,

MANNY: Mo is absolutely correct, Jackie. You just have to get used to a different way of thinking! Bohr taught us, most conclusively, that the way the real world is depends on the way we look at it. So he concluded that there must be complementary ways, even though they may seem logically exclusive of each other. This is called the 'principle of complementarity.' An example of this is the idea of 'wave-particle dualism,' that you mentioned before, Mo. If something is seen as a 'wave,' then it is a wave at the time. If it is seen as a particle, it is a particle then, if the observation is designed to see it so. It is one of the great conceptual discoveries of physics of the twentieth century! My students feel very uneasy with this, especially at first, but, eventually, they get used to it and they stop fidgeting. Then what is your problem, Jackie?

MO: Yes, I agree with you, Manny. Bohr's notion of complementarity is true. It is like an Escher painting.

Manny and Mo were looking as pleased as punch, as they stared at Jackie as though she had lost her senses! Jackie had a look of deep frustration as she gazed out of the window. Beads of sweat were breaking out on her forehead and on her upper lip when she replied,

JACKIE: Look, what you are both trying to say, whoever you may believe taught the lesson, is that there is no objective world at all! You are saying that the way the world is depends on the way you look at it. Let me ask you, then, what would happen to the world if all of the scientists ceased their observations in experimentation? Would the real world disappear? Mo and Manny, I cannot see how you actually can accept this idea, whether or not it came from the great scientist, Niels Bohr? Do you think that the law of gravity did not exist the day before Newton discovered it in the 17th century?

MO: Look, Jackie, let's not get wild about this. It's just that you have to get with this new type of thinking. We have discovered, beyond the shadow of a doubt, that the most elementary form for the laws of Nature is that of a theory of measurement whereby a macroobserver measures the properties of the microelements of matter, where these laws are in the form of a probability calculus. It is called 'quantum mechanics.'[2] Your thinking, Jackie, is the old-fashioned sort that is called 'classical physics' — the mystical idea that there is a world to talk about, scientifically, aside from measurements of it! This type of thing has simply been refuted, once and for all, by the experimental facts of modern physics.

Day and Night

MANNY: Mo is right, Jackie. It is absolutely necessary that you cast off those old ideas and let yourself go! — free to think in any direction at all, so long as you wind up in the camp where the empirical facts support you; that is, in the direction of quantum mechanics! We know now that the laws of matter are in the form of a probability calculus because the elementary things of the material world — the electrons, protons, quarks, etc. — are intrinsically probabilistic entities.

JACKIE: In fact, Manny, the quarks are so probabilistic that it is impossible to see them at all![3] All that you can do is to infer their existence from other effects, and only in the context of a particular model of nuclear matter! It reminds me of the claim that came from a large particle accelerator facility, not long ago, that a new sort of particle was discovered that has no electrical charge, no spin, and no mass. The only way that you know about it is that it leaves a putrid, awful smell where it was — after it is gone — they called this particle a 'pewon.'[4]

Manny was not in the mood for Jackie's attempt at humor at the expense of the large effort that was going on in the physics world of high energy particles. He continued his advice to his rebellious colleague.

MANNY: Look, Jackie, it is absolutely necessary that you learn to accept the truth, if you are going to contribute to the future progress of physics and teach the truth to our students.

MO: Yes, as well as one of the newly discovered views that is being advanced today, that the elementary quanta of the real world are indeterministic, rather than ordered. It has been a very exciting discovery that the basic laws of the world are rooted in indeterminism rather than the old-fashioned idea that the world is ordered.

JACKIE: Why in the world, Mo, are you so certain that what it is that is real is a fundamental uncertainty, so to speak? Isn't this 'certainty' of 'uncertainty' a contradiction in terms? Doesn't it say in the Biblical Scriptures that in the beginning there was chaos, and out of this chaos came order? Maybe what you are saying is that the Biblical account is turned around! Maybe it should have said, instead, that in the beginning there was order, and out of this order there came chaos! The believers in the reality of quantum uncertainty must certainly believe this!

MANNY: Yes, Jackie, we do believe that the laws of Nature are indeed laws of probability. It is a lesson of the facts of 20th century physics of the micro domain of matter, that the laws of Nature, at their irreducible level, are laws of probability, not laws of order![a] Why don't you know this? Everyone else seems to know it! There are many experiments to prove it. A clear example is that of a weak beam of photons passing through a polarizer and then an analyzer. The final beam clearly lacks a total order.

MO: I don't recall this particular example, Manny. Please refresh my memory.

JACKIE: I know the argument, but let's hear it again — to see if we can punch any holes in it!

MANNY: The idea is this. One sends a weak beam of electromagnetic radiation toward a crystal polarizer, say in the positive x-direction, to the right, whereby the axis of polarization is set up in the z-direction (i.e. upward). If the beam is weak enough, we know that it can be considered as one photon at a time passing through the polarizer.[5] Before they enter the polarizer, the photons have all possible polarization directions in the plane that is perpendicular to their direction of motion. (Manny draws the picture of this set-up on the blackboard, as in Fig. 1).

The polarizer then serves as a filter — it allows only those photons to pass that have its direction of polarization, that is to say, the polarizer 'prepares the state' of photons, polarized in the z-direction, that then propagate toward the second polarized crystal, the analyzer.

MO: What happens to the photons that were not polarized in the z-direction, Manny?

JACKIE: Well, these don't have to be photons, my friends. They can just as well be a continuous electromagnetic radiation field, though of very low intensity. The radiation that is polarized in the direction of the polarizing crystal passes through but the part of the radiation that is not polarized this way is either absorbed by the crystal or reflected by it. So far, the discrete photon model is not a unique way of describing the state of affairs, is it Manny?

MANNY: Just be patient, Jackie, and I will get to that part of the experimental results that is unique to a model in terms of photons with undetermined paths. Now if the second polarized crystal (called 'the analyzer') should be oriented with its direction of polarization parallel to that of the first polarizer, all of the prepared photons would pass through it, and if it is oriented perpendicular to it, none of the photons would pass. This is the same result as the one that Jackie would anticipate with her continuous radiation model, that was known to Michael Faraday in the 19th century! But if the analyzer is polarized at 45 degrees from the z-direction, one photon passing at a time would not know which way to go, that is, to transmit through the crystal or to reflect from it! — or to rotate its direction of polarization away from the z-axis.

What we do, then, is to place a photomultiplier between the polarizer crystal and the analyzer crystal, to measure the number of photons and the time sequence in their motion on the way to the analyzer, and then place another photomultiplier on the other side of the analyzer to look at the sequence of photons that had passed through, when its polarization direction was at 45 degrees from the direction of polarization of the entering photons.

What we see in this experiment is that the photons that pass through to the other side of the analyzer do so randomly, that is, they arrive at random times, not consistent with an ordered trajectory, as Newton's deterministic theory of trajectories would anticipate. Also, some (unpredetermined) number of photons would be polarized in the *z*-direction while some other (unpredetermined) number of them would be polarized in the *y*-direction.

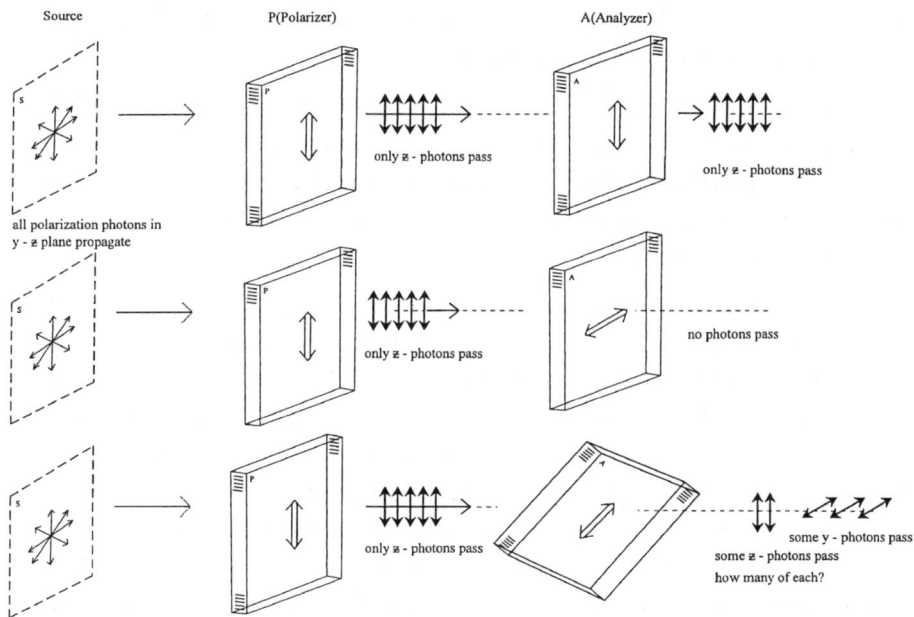

Fig. 1 P Crystals are polarized in the *z*-direction. Radiation propagates in the *x*-direction.
   a) Analyzer crystal polarized in the *z*-direction.
   b) Analyzer crystal polarized in the *y*-direction.
   c) Analyzer crystal polarized at 45° in the *y-z* plane.

MO: Yes, Jackie, this is the same sort of fundamental randomness that one sees in the studies of the absorption of particles from a radioactive material, in triggering a Geiger counter. We see that the pulses occur in a random sequence, like: 1 11 1 1...1 1... ...1... This proves the randomness of Nature, doesn't it Jackie?

JACKIE: No, it doesn't prove anything conclusively, Mo! Your conclusions about reality, based on these experimental results, are only as true as the assumptions about Nature that you started with, in terms of quantum particles. Don't you think it is possible that a different sort of model of reality from the outset, based on entirely different ideas, could give predictions for the same empirical results from an epistemological basis that is fully rooted in order rather than fundamental disorder?

MANNY: No, I am certain that no other physical theory could possibly give these empirical results about elementary disorder, especially if it is a theory based on the concept of an underlying order.

MO: I think that Manny is right, Jackie. No other physical model could give the observed randomness, except for a model in terms of intrinsically disordered elementary particles of matter. It seems clear to me that the inherently uncertain motion is one of the irreducible objective features of the basic building blocks of the material universe.

JACKIE: But how do you know, with such certainty, that this is necessarily the case? Look, my learned colleagues, why are you so certain that what appears to you, for example, as a random sequence of clicks in a Geiger counter, or the random arrival of photons in a photomultiplier detector, is not really a set of ordered peaks, connected coherently in a fully ordered continuous signal? (Jackie goes to the blackboard and draws a diagram, as in Fig. 2).

Do you see, the horizontal line, V, represents the voltage bias that has been set by the experimenter, wherever he/she wishes it to be. Any signal with less voltage than this preset value will not be seen; those with more voltage will be seen. However, as we see in this coherent, even though complicated curve, all of the detected peaks are really part of a single, continuous curve. If the experimenter decides to lower the voltage bias line, more signal would be detected, and usually more noise. Still, the actuality is the full, coherent curve, whatever it is that the experimenter decides to do or not to do. This actual curve is the reality of Nature, and it has nothing to do with intrinsic disorder! Clearly what seems to be elementary disorder in the trajectory of a photon is only the experimenter's (subjective) knowledge about the (objective) curve. To equate the two is to equate knowledge with the object of knowledge, that

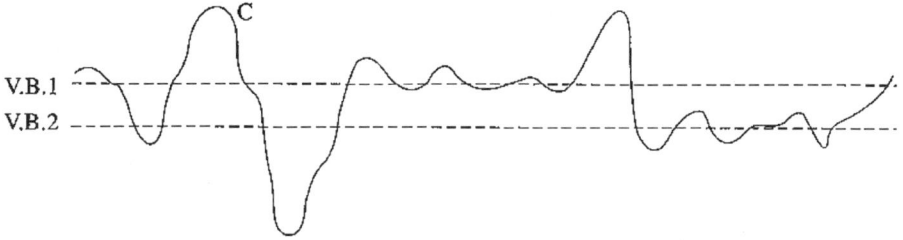

Fig. 2 (a) When the voltage bias of the counter is at V.B.1, clicks are heard when voltages are above VB1 — seemingly at random intervals.
(b) When the voltage bias is at V.B.2, a different seemingly random distribution of clicks are heard. Yet, underlying each set of seemingly random clicks is a predetermined coherent voltage curve, C.

is, it is to equate epistemological conclusions with ontological conclusions. This is clearly false. Certainly, there is more to the entire universe, in any of its domains, from cosmology to elementary particle physics, than a human being's knowledge of it, at any particular time in history! To deny this would be to regress to the pre-Copernican state of knowledge.

Manny ignored Jackie's reference to philosophical concepts in the argument, which he dismissed as having nothing to do with physics, *per se*. He then replied to Jackie's curve.

MANNY: Look, Jackie, this is truly ridiculous. Where could such a coherent curve come from? I don't even see that there is a mathematical function that could represent your curve, thus there couldn't be a law of Nature whose solution is this curve. It is clearly a figment of your imagination and nothing more.

JACKIE: I think that you're letting your own ego control your rational thinking on this problem, Manny. Just because you don't happen to know where the total curve I drew came from, does not mean that there cannot be a law of Nature that predicts such a curve! Why should Nature wait for you to recognize Her language before being one way or another? It may be that the field of mathematics is not yet sufficiently developed by human beings to make us aware of this language to represent true behaviors. Maybe, for example, the law of matter that governs radioactive decay is ordered but highly nonlinear. Perhaps the specific nonlinear behavior simply does not manifest itself in the types of simplistic experiments that we are doing at this stage of physics. That is, we must catch up with the way Nature really is. But whether or not we catch up, Nature will behave in the way that She does, should we understand it or not!

MO: I agree with Jackie on this, Manny. I confess that I was thinking like you, Manny, this morning. But this afternoon, I can see that this is a bit too anthropomorphic. Still, our observations must play an important role in our expressions of the laws of Nature. I do see what you are saying Jackie, that we are looking into an objective world, to learn about the way it is; our ideas do not form a world by the way we look at it.

JACKIE: Yes, Mo, our observations do play an important role in our expressions of the laws of Nature, based on what we see in experimentation. But what we see is not necessarily the whole truth, or even a small part of it. The idea of science is to proceed rationally from our experimentation to the underlying reality of the world. It is the idea of Plato's metaphor of the cave.[6] The ultimate aim in science is to explain Nature, not only to describe Her. The explanation, in turn, is in terms of an underlying reality that is responsible for our observations, that is, the causes for the shadows on the wall of the cave.

MO: Yes, I remember Plato's metaphor. He suggested that the reality outside of the cave, which we initially do not perceive directly, causes shadows on the cave wall, which we do perceive. It is then up to us, according to Plato, to use the ideas that form in our minds, to identify the logical structures in order to deduce something about the reality of the world, from the shadows that we see. This is how Plato saw us approaching some real comprehension of the reality of the world — as little as this comprehension may be at any particular stage of our knowledge, compared to the total objective knowledge of the universe that is there.

MANNY: Don't you see, Mo, in science, the shadows themselves are the only reality. They are the only thing that is truly meaningful for us to talk about. This reality that you talk about, that supposedly creates the shadows on the wall of the cave, is nothing more than metaphysical clap-trap! We have grown up in the 20th century. We have learned from Bohr and Heisenberg that in fact there are only shadows on the wall; there is nothing more to the domain of scientific thinking.

JACKIE: What makes you so sure, Manny, that what we learned from Schrödinger and Einstein, about the existence of a reality as underlying what we see in physics experimentation, is not closer to the truth, and that indeed, such a point of view would be more helpful toward the shaping of physics research in the future?[7]

MO: I am confused! Perhaps for the time being, we can say that we have clearly defined our positions. But I will certainly have to do more thinking about these differences of opinion in physics, between now and next Tuesday, when we meet again. In the meantime, what am I supposed to tell my students? They like to hear that everything is settled, that the scholars know the ways of Nature. It makes them feel secure about their studies! Tell me, Manny and Jackie, what should I tell them?

MANNY and JACKIE: (in unison) Present the facts of the arguments to them, on both sides and with equal vigor, then tell them to think for themselves and try to come to their own conclusions!

MO: All right, we've had enough debate today to mull over until next time. Let's carry on from this point at our next Tuesday meeting.

Before leaving the Conference Room, Manny, Mo and Jackie were caught by the beauty unfolding through the western window. An exceptionally large orange sun was setting over the ever-rushing river, just at the edge of the campus. As the orange reflection shone on their faces, they were mesmerized by the view. Manny had a sudden thought that perhaps Plato was right after all! Perhaps there was more to the universe than the sun and the river, beautiful as they are. How much more beauty is there, yet unseen! Jackie and Mo knew that Manny had that thought at that moment, by the wonderment that they saw on his face.

WEEK 2

WHAT IS LIGHT?

It was four o'clock in the afternoon again, and Manny, Mo and Jackie were pouring their brewed coffee. Everyone had gone home and it was deliciously quiet in the Physics Department Conference Room. The sun's rays were especially bright as they penetrated the west window from a point in the sky not too far above the horizon, along the rushing Willamette. It was getting to be late fall and the days were becoming noticeably shorter. Jackie put her cup down, after a long sip, and started to open her mind to her colleagues.

JACKIE: Our discussion last week about single photons and wave-particle dualism made me think about a more primitive concept altogether, that of light in the most general terms. I'm not so sure that I have any understanding at all about what light is supposed to be, according to anyone!

MANNY: What do you mean, Jackie? If there is anything that we do understand in physics, it is the concept of light!

MO: I'm not so sure that we do, Manny. I recall reading Einstein's recollection. In his later years he said that when he was young he felt that while he didn't yet understand the nature of matter, he was happy to say that at least he did understand what light was all about. But when he became older, after a lifetime of studying matter and radiation in physics, he said that if there was one thing he didn't yet understand, it was light!

Are you sure, Manny, that you really understand the nature of light? Einstein seems to have been saying, after a lifetime of study, that 'there is more to light than meets the eye!'

After acknowledging Mo's pun of the day (there was usually at least one of them, although one never knew when or where it would appear!) the three colleagues started to gaze out of the eastern window, where they saw a panoramic view of misty cloud cover rolling over the Cascade mountains, silently starting to fill their valley and threatening to snuff out the brightness of the late afternoon sun.

JACKIE: We won't have sunshine in this room for more than another hour, so let's get on with it. I find it difficult to discuss the true nature of light under the artificiality of the light produced with neon bulbs!

MO: You are strange, Jackie. You don't seem to be able to think straight without authentic sunshine! Why is this?

JACKIE: It is something between me and Nature, Mo. Look, the sun that we see over there may in fact be gone at this moment. It may have just snuffed out. We are seeing the sun the way it was eight minutes ago. How do we know that it is still there, right now?

MANNY: Sure it was discovered centuries ago that light propagates at a finite speed. What is so perplexing about that Jackie?

JACKIE: What I am perplexed about is this, Manny. What, exactly, is it that is propagating from the sun to us, that we call by the name, 'light?' Clearly, something happens in the sun that is physical, and then, eight minutes later, our eyes respond with an electrical signal that is sent to the optical sections of our brains, that gives us the sensation of light. But, aside from our physical brain reactions, exactly what is this 'light' that is supposed to have propagated to us from the sun, 93 million miles away?

MO: We all know, Jackie, since the discoveries by Maxwell and Faraday in the 19th century, that light is an electromagnetic wave disturbance, propagating from there to here.

JACKIE: I know the jargon, Mo. But I am asking what the 'it' is that is propagating to us. I mean, if we talk about a water wave, say a breaker on the ocean front, we are describing a sequence of increasing and decreasing amounts of water, moving along the surface of the ocean. Analogously, exactly what is light a wave of?

MO: We used to have an ether theory. Newton believed that the corpuscles of light propagate by displacing ether atoms. Maxwell believed that there is, similarly, a background that light moves through. It is the ether — in Maxwell's view, a lattice of ether atoms. Light is then supposed to be a disturbance of this ether, whereby its atoms vibrate in a particular way, transverse to the direction of propagation of the light, as was discovered by Fresnel.

But then Einstein showed that there is no need for an ether to describe light; this conclusion followed from his theory of special relativity. In his theory, then, light is a manifestation of the electromagnetic free field — a vibration that is a thing-in-itself!

MANNY: Recall, Mo, that it was Faraday's thought that the electromagnetic field, generally, is a continuously distributed potentiality.[b] This potentiality, as in Aristotle, does not manifest itself until it is actualized. In Faraday's theory, the actualization happens when a test body appears somewhere in the field of potentiality, wherever that is. The test body is electrically charged material, that is, anything that would respond to an electromagnetic force, such as a television antenna responding to the signal of a TV program. According to Faraday's view, it is this actualization of the electromagnetic field that is the reality of light. For example, the electrical charges on the backs of your eyes, in the retinas, are the test bodies whose responses to your brain give you the sensation of light.

MO: This idea seems similar to the basis of the quantum theory as a theory of measurement, doesn't it, Jackie? Isn't this the idea that Manny is trying to support? You see, Jackie, the test body, in Faraday's theory, plays the role of the measuring apparatus of the quantum theory — a device that actualizes the interaction with the micromatter, just as the test body of Faraday's theory actualizes the presence of the electromagnetic field.

JACKIE: Yes, there may be a similarity in this way, but there are major differences between Faraday's view and that of the quantum theory. One of the differences is that in the Faraday field theory there is no difference between macrovariables and microvariables. Also, there is no 'collapse of the wave packet' when the actualization takes place, regarding the interaction between the test body and the field of force.

MO: Still, the Faraday idea does have in it the field as a potentiality, just as the possible states of matter in the quantum theory are its potentiality. In both cases there must be an actualization in order to create the reality. Thus, the reality of light is created when the test body actualizes the potentiality of the electromagnetic field of force, just as the reality of the physical property of micromatter is created when the measurement is carried out by a macroapparatus. This is what Bohr was trying to teach us, wasn't it, Manny?

MANNY: I agree with Jackie, that there are more differences than there are similarities between Faraday's theory of light and the photon model of light, according to the quantum theory. With the quantum view, it is important to assert that there is no field of potentiality to underlie the measurement. The laws of Nature — i.e. the laws of quantum mechanics — are laws of measurement. With this idea, light cannot be a thing-in-itself! Rather it is the result of a particular sort of measurement. But the law of measurement is the law of Nature, there is no underlying field that is basic, as there is in the Faraday field theory. Do you see the difference, Mo?

JACKIE: I still don't see what the object is that you are discussing when you refer to light. When you say that 'it' propagates from here to there, what is the 'it?'

MANNY: An 'it' of light is a collection of physical attributes of an atom of light that we call 'photon.' What I have been saying is that these attributes of the photon are not properties of a thing-in-itself. Instead, they are intimately related to our measurements. This was a very important advance in our knowledge of matter as well as light, and was taught by Bohr and his colleagues in Copenhagen. Thus, we have found that the physical attributes of photons are that they are a collection of waves, when we look for interference effects in experiments on light, and they are a collection of discrete particles when we look for particle-like features of the photons, as in the photoelectric effect or the Compton effect, dealing with the scattering of electrons by photons. This is the first example of wave-particle dualism that was discovered in the history of modern physics, by Einstein himself! — the prime opponent of the Copenhagen school! But this is the reality of light, do you see this, Jackie?

JACKIE: But the logic of fields is entirely different than the logic of particles. How is it that something that may be described as a collection of localized particles may also be described as a collection of continuously distributed waves — just because we choose to look at it in one way or the other? Look, Manny, if I should hold two pieces of chalk in one hand and three in the other, I must have a total of five pieces of chalk — there is no other answer! But if I have two waves added to three waves, I can get anything at all for their sum, depending on their relative phases.

Anybody can see this difference. I recall when I was a small child, when I was sitting on the beach, very close to the shore, I was watching a ripple of the ocean roll up on the wet sand — too wet to absorb the water. The ripple then started to move down again toward the ocean. Soon, another ripple came up to the first one and it would annihilate it or strengthen it, depending on the timing of when the two ripples met. I recall noticing then that one ripple plus one ripple could be no ripples — or anything else. This was an interesting demonstration of the way waves combine, showing how different it is than the way discrete things add up. Though I was only four years old then, I could see that the logic of adding continuous waves and the logic of adding discrete things is entirely different. I could see this, without even knowing the meaning of the word 'logic!'

So you see, Manny, it just doesn't make sense to me to say that light is a collection of discrete things if I look at them in this way and also a collection of continuous waves — something quite opposite — if I look at them in that way! It is illogical — positivistic or not![1]

MO: Is it possible, Manny, that one might say that light is really a collection of waves, but that it only appears as localized particles under very special conditions of observation? — or vice versa? For example, if one should move quickly past a screen door, it would appear to be a continuous gray smear. But a view, close up, would reveal that it is really a grid of discrete wires.

MANNY: No, this is not the idea we are talking about with the quantum view. There is no such reality here that is supposed to underlie the measurement. The principle of complementarity that Bohr found to be basic to the laws of micromatter, and perhaps in many more settings, says that the only reality is the measurement itself. Why can't you understand this? The empirical evidence reveals, beyond the shadow of a doubt, that under some sorts of experimental conditions light 'is' a collection of discrete, localized particles, like pebbles without inertial mass, but certainly with momentum and energy. But under other sorts of observation, light 'is' a collection of continuous waves. Whether or not this may seem to insult your common sense, it is a fact of Nature that we all must accept.

JACKIE: Then what you are saying, Manny, is that there really is no 'it' to talk about, if we are talking about light, *per se*. Rather, light is a response of a human brain, or other instruments, to an electromagnetic field of potential force. The way light seems to be is exactly the way we see it — nothing less and nothing more than this immediate response! Is this correct, Manny?

MANNY: Yes, that is the idea.

JACKIE: But I still don't understand the claim that light travels at 30 billion centimeters per second! I mean, what is it that is supposed to be traveling at this speed?

MANNY: It is a potentiality that travels at this speed, from its source, say the sun, to the point of observation, such as your eyes.

JACKIE: But you are claiming, then, that before the light gets to my eyes, it was traveling toward me from the sun, as a potentiality. It was then an 'it' before my powers of observation came into the picture. Doesn't this contradict what you said before, that light is not a thing-in-itself, but rather only an actualization of a potential?

MO: I can certainly understand now why Einstein found it difficult to understand what light is all about, especially in terms of the modern quantum theory.

JACKIE: One other peculiarity that is very difficult to understand about light is this. When one is describing a single photon, mathematically, that is, a quantum of light with a single frequency, say yellow light from a sodium lamp, then this photon seems to be located everywhere in space! I mean, the scientific description of this yellow light is that it is everywhere in space, rather than being in a single place. If this is a true description, and if this light is moving at the speed of light, where is it going to and where did it come from?

MANNY: Well, actually, when we detect photons in our experimentation we never see a single one of them. They always appear in a wave packet, which is a distribution of photons with different frequencies, though close to a central frequency, such as that of the yellow light. The description of the wave packet, in turn, is localized in space, as one would expect for a localized object.

JACKIE: But we were talking about a single photon, not the way a bunch of them travels together, like a pack of wolves! According to what you have said, Manny, it is meaningless to refer to a single photon. In fact, last week, we talked about a thought experiment in which a beam of light is so weak that it corresponded to a single photon at a time entering some detecting apparatus. In the context of what you have just said, how could that be sensible?

MANNY: I agree with you, Jackie, that the description of the single photon implies that it is everywhere in space at once and that it is traveling at the speed of light. But our macroapparatuses don't know anything about the lack of logic in this, until we detect the light. At that time it becomes logical! It suddenly appears for us where we look for it, and it is nowhere else in space any longer.

MO: I do find that hard to swallow, Manny. I agree with Jackie that science should be based on a logical structure — that is to say, that the laws of the universe are based on a certain order. Your idea of light, Manny, seems to me to be as illogical as it can possibly be.

Recall Einstein's remark: "God may be subtle, but He is not malicious." I think that what he was trying to say was that the laws of the universe are logically simple. And to be logically simple, the laws must not lack in any order.

MANNY: Why do you think, Mo, that logic is so holy? Did it ever occur to you and Jackie that perhaps the universe is not really based fully on logically ordered laws? After all, these rules of logic are the creations of human beings. Isn't it possible that God had something else in mind, not these rules?

JACKIE: Isn't it possible, my friend, that your rejection of logic from the total representation of the laws of Nature is due to your frustration with some of the problems of illogic that arise in your approach? Is it possible that you are just trying to sweep your troubles under the rug?

MANNY: No, that is not the case! What I believe is that we scientists should be as open-minded as possible if we are to accurately represent the ways of Nature, objectively. Just because you and Mo insist on a logical basis for the laws of Nature, as has been assumed by scientists and philosophers since the ancient period of Greece, does not make this idea necessarily true! Isn't it possible that, in fact, $2 + 3 = 5$ and $2 + 3 = 26$ are both true propositions when we are relating to the true, objective features of Nature? Why should Nature believe in the rules of logic, just because we human beings invented them?

MO: This reminds me of the famous anecdote about President Roosevelt, when he was trying to listen to opposing arguments of two senators. When one of them completed his case, the president answered, "you know, you are right." Then the second senator presented his case, which was opposite to the first senator's, whereupon the president answered him, "you know, you are right." The president's wife, who was in the room at the time, then responded, "Franklin, how can both of these gentlemen be right if they were each presenting logically opposite points of view?" The president then responded, "you know, Eleanor, you are right."

Isn't it possible, Jackie, that the same sort of lack of logic is true of the natural world, at least in part?

JACKIE: That anecdote, Mo, really goes back to Jewish folklore from Eastern Europe. It is in a story about a rabbi and his congregants, written by Sholem Aleichem. But it was meant to be a joke, certainly not to be taken seriously!

I do believe, Mo, that the way the real world is — in terms of its fundamental laws — is based on order, not on disorder. This order, in turn, is expressible in terms of a logical pattern — whether or not we have developed our intelligence far enough to discover this order! It may be that there are different sorts of logic to be utilized in the language of Nature — single-valued logic, modal logics that are multivalued,[2] or whatever other sorts of logic we don't know about yet. But some sort of logic is necessary to express the order that underlies the objective truths of the universe.

MO: I'm not sure now that Manny is wrong about this, Jackie. You are assuming that Nature's laws are logically ordered. Maybe this is not so. Perhaps the randomness that we discussed last week is indeed fundamental in Nature, thus refuting your idea of total order.

When we parted last week, I felt myself swerving in your direction, Jackie. But now I'm not so sure that Manny is wrong. In any case, you both present convincing arguments — one based on the idea that the universe is characterized by a total order and the other on the idea that disorder is more basic to the true nature of the universe.

I think that we should leave it now to mull over until next Tuesday. Perhaps we never will understand light!

Manny, Mo and Jackie walked toward the large oak door of the Conference Room with elated spirits — even though they hadn't yet solved their problem of light. Manny was feeling a bit smug about his illusion that he had conclusively convinced Mo and Jackie that disorder is a fundamental ingredient of the universe. But Jackie was simply feeling frustrated because she still believed, with all of her soul, that the universe is based on order — just as the early Greeks did. However, she realized then that her belief in a total order was based on a faith in this idea, perhaps not shared by most of her colleagues in the physics profession, at least at this point in time.

The large oak door was clicked shut. Manny, Mo and Jackie were on their respective paths to their academic responsibilities. As they parted, they called to each other with twinkles in their eyes, "until next Tuesday!"

RELATIVITY AND THE TWIN PARADOX

A brilliant white cap of snow was covering the tops of the mountains, and the sky was clear blue. Mo was waiting for Manny and Jackie, who were detained at the faculty meeting that he had been able to avoid this time. He was thinking about the controversy on whether the world might really be based on fundamental uncertainty or whether it is rooted in order. The experience of the meeting of the Physics faculty would certainly have gone in the direction of uncertainty! But human relations are much more complicated than problems of physics!

After all, Mo pondered, isn't a scientist's credo to search for the cause-effect relations that underlie the reality of the world, and aren't these expressions of order? If Manny wants to find the laws that control the randomness of the world, aren't these laws, themselves, expressions of some sort of order?

The sound of pouring brewed coffee and its aroma, at the other end of the Conference Room, broke into Mo's meditative state of mind.

MO: I know that we all believe in the democratic principles. But it does not always work — that a vote can decide on the truth of a proposition.

JACKIE: We see this all the time, Mo, regarding the breakdown in terms of physics ideas whose truth is based on consensus of opinion.

MANNY: I wonder if we applied this to the beliefs of our first-graders, would we get their decision on a truth, such as the claim that $2 + 3 = 5$? If 80% of the first graders vote false, 15% vote true and 5% are undecided, would this mean that 2 plus 3 is not equal to 5?

JACKIE: I'm glad that you see it this way, Manny. I am often reminded of the case in physics regarding the so-called twin paradox of relativity theory. Everyone seems to agree that there is a truth where there is really a logical paradox. Obviously, a paradox cannot be a scientific truth!

MANNY: What do you mean, Jackie? There is nothing wrong with relativity theory! The physics profession decided a long time ago that there is this time effect according to relativity theory, and there is a consensus on it.

MO: According to what you said about the first-graders, Manny, the consensus doesn't make it right!

MANNY: If it were not true, there wouldn't be a consensus of all of the learned physicists about this question!

JACKIE: Let's look at this twin paradox situation, Manny, in a rational way. How can a paradox be a fact of Nature, based on a consensus or not?

MO: What is the paradox, Jackie? I've forgotten exactly why there is supposed to be a paradox coming out of Einstein's theory of relativity. I do teach my students that it is a prediction of relativity theory that time will pass slower for an object that moves relative to a fixed observer, compared with the fixed observer's time. But where is the logical paradox in this admittedly peculiar situation?

JACKIE: There is a logical paradox because motion is subjective; that is, it is relative to whoever is the observer. Suppose that we have twin brothers, Paul and Peter, both born at the same time and therefore being the same age, mentally and biologically. Suppose that Paul moves eastward from Peter. Then one could equally say that it is Peter who is moving westward from Paul. Now if the motion predicts a slowdown of the aging process, then Paul would age at a slower rate than Peter, from Peter's view, but it would also be true that Peter would age at a slower rate, from Paul's view. Thus, according to Peter, when Paul takes a trip, first going eastward, then turning around and returning, westward, he would be younger than Peter, biologically, when he gets home. In fact, the faster Paul travels relative to Peter, the greater would be their age difference when he returns home. Before Paul embarks on the trip, both twins are, say 22 years old, but after he returns home, Paul may be 22 years and one week old, while Peter has reached the age of 98 — a doddering old man with no teeth, bent over, no hair and a bit senile; he may have already forgotten that he has a twin brother, Paul!

MO: When Paul changes his direction of motion, from eastward to westward, when he turns around to return home, would he then age at a greater rate so that when he meets Peter he would again be the same age?

JACKIE: No, the age difference, according to this current interpretation of the formula from relativity theory, that is, the Lorentz transformation of the time measure, depends only on the magnitude of the velocity, not on its direction.

MO: Where is the logical paradox, Jackie? It is, of course, an unexpected result, that because of his motion, Paul would age less then Peter. But this is not a logical paradox in itself!

JACKIE: The paradox is created by the fact that motion, which is supposed to be the cause of an absolute physical effect, is relative. I mean this, Mo. If Paul moves relative to Peter, first eastward and then westward to return home, then it is equally true to say that Peter is in motion relative to Paul, first westward than then eastward, to return home! Thus, if we came to the conclusion that Paul would be younger than Peter after they meet again, it is equally true that Peter would be younger than Paul after the trip is completed. That is, the conclusion would be reached that Peter is both older and younger than Paul, biologically! This is a logical paradox. Thus, it is unacceptable as a scientific statement!

MANNY: Well, there really isn't a paradox because there is a bona fide aging effect due to motion. We've never seen it in regard to human beings because we haven't been able to move at speeds close to the speed of light. The effect is always there, but you just wouldn't see it because it would be too small, according to the formula of special relativity theory — unless one could travel at speeds close to the speed of light. However, there are all sorts of real experiments that can be done where we would see this prediction of Einstein's theory of relativity.

JACKIE: What prediction of relativity theory, Manny? I just showed you that if the theory of relativity does in fact predict this effect, it would lead to a logical paradox. Such a conclusion would then nullify the theory of relativity as a bona fide scientific theory! Your claim about experimental facts is beside the point! That is to say, if the theory of relativity is a scientific truth, which I believe to be the case, then its formulas that entail time measures must be interpreted differently than you are doing — if we are to avoid the paradox!

MANNY: I still insist, Jackie, that the experimental facts back up the theory's prediction of an asymmetric aging effect. Let me recall one of these for you, Jackie. There is the example of the data on mu mesons from outer space that come with the cosmic rays.[c] When it is at rest, the mu meson has a mean lifetime of about a millionth of a second, before it decays into an electron and two neutrinos. But at the speed that mu mesons are traveling and the distance that they come from, they would never get here in a millionth of a second! To get here, they must live much longer than this when they are in flight. In fact, the formula from relativity theory, that is, the Lorentz transformation for the comparison of time measures in relatively moving frames of reference, gives just the right increase in lifetime to account for the observed mu mesons that originate in these cosmic rays.

MO: Before we go on with this, Manny, I see something very wrong in this particular example. Look, the time we talk about in special relativity theory, that we call by the name, 't,' is a parameter that measures the evolution of the spatial location of a single particle as it traces out its trajectory; that is, the time change corresponds to a change of location of the particle.

MANNY: Yes, I agree with this. But we also use the parameter 't' in the expression for the decay rate of the mu meson. Then what is the problem, Mo?

MO: It is this. The mean lifetime of the mu meson, or any decaying matter, refers to an average of a very large number of particles. It does not refer to the trajectory of a single one of them! A single mu meson, out of a bunch of them that come to us in cosmic rays, may have a lifetime that is much longer than a millionth of a second or much shorter! This sort of 'time' measure, that is only in reference to an average time measure for a bunch of mu mesons, is a totally different concept than the time 't' referred to in the formula of relativity theory, for the trajectory of a single thing. For this reason, Manny, I don't believe that the relativistic time parameter, as a measure of motion of a single particle, and the contraction of this time measure in the moving frame of reference, according to the formula of relativity theory, that is, the Lorentz transformation for the time measure, has anything to do with the time measure that you are talking about, regarding the mu meson's lifetime.

JACKIE: Beside this, Manny, we never actually measure the time that the mu meson exists, before it decays to the electron and two neutrinos. What we do measure in this regard, in our experimentation, is related directly to the rate of change of the number of mu mesons in time. One then extrapolates from this measure of a rate to the actual number of mu mesons that have decayed. It is the number, and not the rate of decay, that relates directly to the average lifetime of the mu meson.

We normally expect to find that the measured rate would be different in different frames of reference — because a rate is a number per time, and the time measure depends on the reference frame, according to relativity theory. It is analogous to the Doppler effect, where one sees, for example, that a yellow line in the visible spectrum, say from a sodium lamp, looks red if the sodium lamp is moving away from the observer. Of course, the observer really sees the line as red, but the mechanism that is creating this radiation in the lamp is the mechanism to produce the frequency of a yellow line. That is to say, the real frequency is yellow, even though we see it as red because of the relative motion.

MANNY: What do you mean, Jackie, by the 'real frequency?' The only reality, as we discussed last week, is what we measure! If the frequency of light is measured (or seen) to be yellow, that's what it is. If it is seen to be red, that's what it is. There is no meaning to 'real' other than the measurement, at the time when it is carried out.

JACKIE: What I mean by 'real frequency,' Manny, is the rate of vibrations of the light that is produced by some mechanism. This cause of the vibration has nothing to do with what you or I happen to see or don't see! In physics jargon, this is called the 'proper frequency.' To measure its value you would have to be at rest with respect to the cause of this vibration — the atoms or molecules that are emitting the light. This is because of the fact that if you happen to be moving with respect to these atoms, or if they are moving with respect to you, you would not respond to the 'proper frequency.' That is, your motion with respect to the source of the light would give you the illusion, from your measurement, that the source of the light is causing it to vibrate at the red frequency, whereas you would find out, by removing the relative motion (either by stopping your motion with respect to the source, or by applying the correct mathematical transformation) just what the proper frequency is, caused by the source of the emission of the light. You would then find out that it is yellow and not red! The illusion that I mentioned is called the Doppler effect. It was first discovered in regard to sound vibrations and later discovered in regard to light.

MO: Coming back to the mu meson, Manny, it seems to me that what Jackie said is true, that the reason that we don't measure the real frequency of the radiation is that what we are measuring is a rate — a number per time — and the 'per time' part is frame-dependent, because time measure,[d] according to relativity theory, is dependent on the reference frame in which we make the measurement. Thus, one should expect to measure a different rate of decay of the mu meson in a moving frame compared to this measure in a frame of reference that is at rest with respect to the meson.

If you really want to know if a bunch of decaying mu mesons is younger because of its motion relative to an observer, compared with its age if it would have been at rest with respect to the observer, you would have to do an experiment like the following: Start with two boxes, each containing the same number of mu mesons, say N. Take one of these boxes on a round trip journey at high speed. Upon returning to the reference frame of the box that stayed home, measure the activity in each of the boxes. If the box that stayed at home has $N/2$ mu mesons left in it, and the box that took the round trip journey has $3N/4$ of its mu mesons left, then the box that took the journey would be younger than the other box. Such an experiment did not involve the measure of rates of decay. It involved a direct measure of activity, which in turn gives the absolute number of mesons that had decayed. This relates directly to an aging.

JACKIE: As far as I know, no such experiment has ever been done. But even if such an experiment were to be carried out, and if it was then discovered that there is such an asymmetric aging effect as you anticipate, Manny, this would not mean that the theory of relativity predicted it! What it would mean is that there must have been some sort of force, that we were unaware of, that caused the mesons in one of the boxes to slow down their depletion, relative to the depletion of mesons in the other box. The physicist would then be obligated to look for such a new force, in order to explain the observed effect. But the theory of relativity, *per se*, does not predict this effect, one way or the other! All that it does say about this is that if such a physical cause-effect relation should exist, it must follow from a law of Nature that obeys the principle of relativity — that its expressions are in one-to-one correspondence in all possible frames of reference, according to any particular observer's view, that is, that this law of Nature must be fully objective, independent of the reference frame.

MANNY: I know what you are saying, Jackie, about the reality of physical phenomena, independent of observations. But Bohr has taught us that this is not correct, if you leave out the measurement of the phenomenon. And you, Mo, are not right about the comparison with the Doppler effect. The mu meson lives longer when it is in flight, compared with a stationary meson, according to the formulas of special relativity theory. This is an empirical fact of Nature. No amount of theorizing can change this scientific fact!

JACKIE: But what about the paradox, Manny? Why can't we equally say that the mu meson that you formerly called stationary, say mu (1), is the one that is moving relative to the other, mu (2), that you formerly called moving. In this case it would be mu (1) that would be longer-lived, rather than mu (2). But according to the principle of relativity, the physics from both frames must be in one-to-one correspondence, that is, the claim from each mu meson's reference frame would be true! We would then be concluding that mu (1) is both longer-lived and shorter-lived than mu (2)! This is a genuine paradox, isn't it, Manny? How do you get out of it and still maintain the theory of relativity as a true theory of Nature?

MANNY: I don't believe that there is a real paradox because the empirical facts support the prediction of the aging effect according to the formula of relativity theory. Theorizing means nothing to me! All that counts is whether or not the empirical facts are in agreement with the theory!

Coming back to the problem, Mo, the situation is not really symmetric after all. This is because one of the twins accelerates relative to the other, and the acceleration is indeed absolute, not relative. That is, while the velocity of one of the brothers relative to the other is strictly relative and therefore symmetric, the acceleration is not because it entails a force that is experienced by one of the twins and not the other. It happens when he turns around in his trip to rejoin his twin brother, as well as the short acceleration times when Paul leaves Peter's reference frame and when he returns to Peter's reference frame. For the sake of argument, let's ignore the latter times and just look at the 'turn-around time' of acceleration of Paul's car.

JACKIE: You are confusing cause and effect, Manny. Acceleration is a 'kinematic' variable, it is descriptive of the motion. It is just as relative as the velocity. If Paul accelerates relative to Peter in his turn-around from the eastward motion to westward motion, to return home, from Peter's view, it is just as true to say that from Paul's view, it is Peter who is accelerating in his turn-around, from the westward motion to an eastward motion, to return home. The relativity of motion, *per se*, is a lesson that Galileo taught us, many centuries ago![1]

MO: Physicists say that while velocity is a relative variable of motion, acceleration is not so because a force is involved, say in turning Paul's car around to change his direction for his return journey to Peter. For example, when we take off in an airplane, leaving the ground in an accelerated motion, or if we suddenly ascend in an elevator from rest, we feel the acceleration in our stomachs. The stationary viewer of this motion does not experience the same tingling sensation in his stomach!

JACKIE: But, Mo, it is a force that is the cause of the acceleration in your examples. Acceleration itself is an effect of this cause. The question is then: does the force exerted by the engine of Paul's car, in turning it around (the observed acceleration effect) also cause other physical effects, such as the slowing down of Paul's biological cell division, so as to physically age him differently, or does this force of the engine cause a clock in Paul's car to slow down its hands that read the time, compared with the reading of a clock in Peter's reference frame? As far as I know, there is no evidence for such a cause-effect relation between the force that causes a body to accelerate and the physical aging of this body!

MANNY: All that counts, Jackie, is that the empirical facts fit the predictions of the formulas that come from our physics!

MO: I don't agree, Manny, that the empirical facts are all that we need in order to explain physical phenomena. The philosophy of science teaches us that while it is necessary for a scientific theory to predict the empirical facts, equally importantly, this is not sufficient to establish the truth of the theory! For the theory to be true, it must also be logically consistent and it must predict unique consequences for unique physical situations. For example, what if some scientific theory predicted an elliptical orbit of Mars, but at the same time, without altering the physics, it also predicted that the orbit of Mars is circular. Such a theory would be unacceptable, even though one of its answers is correct! Or suppose that the same theory predicted that the eccentricity of the ellipse of Mars' orbit has a particular value, and that this value is in agreement with the observations, say to one part in a million. But suppose that the same theory predicted a different value for this eccentricity, without in any way altering the physics, that is in disagreement with the data. Such a theory would not be acceptable! — even though one of its predictions was correct!

Jackie, you mentioned the principle of relativity as the underlying assertion of the theory of relativity. What, exactly, does this principle say?

JACKIE: This principle is the main idea that underlies the theory of relativity — whether we are talking about special relativity or general relativity. The principle says that the laws of Nature must have the same form in all possible reference frames, from the view of any particular observer. That is to say, if a particular phenomenon is explained with a particular law of Nature, let us say in the observer's frame of reference, such as his own laboratory, then if this same person should study the same phenomenon in a different reference frame, let us say a similar laboratory on a comet, flying past Earth at a great speed, then the law of Nature for this phenomenon expressed in the language of space and time coordinates (or any other language!) that is appropriate to the comet's reference frame, should be in one-to-one correspondence with the expression of the law for the same phenomenon in terms of the space and time coordinates of the observer's own reference frame — those of his own laboratory!

MO: This is equivalent to saying that the laws of Nature must be totally objective — independent of any reference frame — if they are bona fide laws of the universe. This seems to me to be a tautology! For, how could a law be a law if it were not totally objective? — by definition of the word, 'law.' It seems, then, that Einstein's theory of relativity, based on this principle, is not a scientific statement, because it is tautological! Further, its assertion of objectivity is not really new — it is as old as the history of science itself!

JACKIE: You are not right about this because when the theory is expressed in terms of specifics, as it was in Einstein, it led to the opening up of a great deal of new physics!

MANNY: How does all of this philosophizing about Einstein's theory of relativity relate to the paradox you claim occurs regarding the twins? And just how do you propose to get out of the paradox while holding on to the theory of relativity?

JACKIE: I cannot tell you how pleased I am, Manny, to see that you are interested in hearing about the other side of the subject that, before today, you thought you understood so well! In answer to Mo's question, I agree that if the principle of relativity were really a tautology, it would be a necessary truth, and therefore not subject to refutation — thus it could not be called a 'scientific statement' at all. But the theory of relativity is science because its principle of relativity entails two tacit assumptions that are indeed contingent on Nature. Thus, it is a theory that is indeed refutable, and therefore its predictions are scientific truths — if they can be verified.

MO: What is the difference between a necessary truth and a scientific truth, Jackie?

MANNY: I can answer that one, Mo. A necessary truth is an assertion that cannot be anything else. It is built-in! For example, when we say that $2 + 3 = 5$, we are stating a conclusion that cannot be anything else, given the definition of the integers and the arithmetic operation $+$. This conclusion about the sum of 3 and 2 is not refutable. But a scientific truth is refutable because it is contingent on Nature. For example, Newton's second law of motion, $F = ma$, means that a force F, that is external to a body whose inertial mass is m, will cause it to accelerate at the rate, a, in a linear fashion (e.g., doubling the magnitude of the force doubles the magnitude of the acceleration, etc.).

However, Newton's law is not necessarily true. It is contingent on the way Nature is. That is, it is refutable. Indeed, this law was refuted in the 20th century by the full form of Einstein's theory of general relativity, even though Newton's second law of motion remained a good mathematical approximation for its counterpart of the Einstein theory, under special circumstances where it was found in the past to be empirically valid (for two and a half centuries of tests!).

What are the specific tacit assumptions that make the theory of relativity a bona fide scientific theory, Jackie?

JACKIE: The first tacit assumption is that there are laws of Nature in the first place. This is the claim that for every physical effect in the world, there is a logically connected cause. This is sometimes called 'the principle of total causality.'

MANNY: I don't believe that this principle is a necessary truth! In fact I don't believe that the principle of total causality is true at all.

JACKIE: I know that you don't believe it, Manny. But I do believe that it is the credo of a scientist — whose job is to search for the reasons (causes) for the natural phenomena of the universe (effects). That is to say, the object of the scientist's study is the discovery of the cause-effect relations that explain the physical workings of the entire universe — from elementary particle domains to that of cosmology. If the object of the scientists' investigations is the full set of cause-effect relations, then, as a scientist, he/she must believe that they exist! That is, the cause-effect relations of the universe are the scientists' raison d'être!

MO: I agree with Manny, Jackie, that the principle of total causality is not necessarily true, and that he has a perfect right, as a bona fide scientist, to disbelieve it! You may have faith in this principle's truth value, but you cannot prove it to be true! I think that you probably feel that any scientist who does not believe this principle may still be a scientist, but an incomplete one! That is, an incomplete scientist is one who cannot go as far along the road of explanation as possible.

JACKIE: Yes, Mo, I do agree with what you have said — no insult intended! It is my definition of 'scientist.' Let me get on, then, with the second tacit assumption that underlies Einstein's principle of relativity. This is the assumption that Nature may be comprehended by us and expressed in a precise way. Perhaps it is arrogant for human beings to believe that, from our infinitesimal corner of the universe, we may actually comprehend something about the entire universe, regarding its fundamental explanation. However, I believe that the history of science attests to this fact, that we have indeed made some progress in our history — little as our accumulation of objective knowledge has been to this point in time, compared with all that there is to understand! But I believe it has at least been finite and not zero! In any case, I have faith that this is true!

The mathematical language that we have invented for the purpose of expressing this conceptual knowledge of the laws of Nature is where space and time come into the picture, in the practical expression of the theory of relativity. What we have discovered, since the initial discoveries by Pythagoras and then Galileo, is that the language of the laws of Nature is a mathematical language, and that this is most readily expressed in terms of space and time measures. What was discovered by Einstein was that if we are to insist that the forms of the laws of Nature remain objective, that is, independent of any frame of reference in which they are expressed, then it is necessary to translate the language of space and time measures from one reference frame to any other, in such a way that there are scale changes in the expressions of the laws in the different reference frames. The expression of a law of Nature, with the language of space and time measures, implies that these measures must contract or expand when we express them in the different, relatively moving reference frames.

MO: If there must be a contraction of time measure in the moving frame, does this mean that physical objects would age at a slower rate in that frame, relative to the stationary frame of the observer?

JACKIE: I can see how anyone can fall into this trap, but it doesn't mean that at all! What the scale change in time does mean is that the time measure must be altered, not that the physical duration itself is altered! For example, looking at a moving clock, we may have to put six numbers on its face, for one complete revolution, rather than the twelve numbers in our own reference frame — in order to correctly describe the law of Nature in the moving frame. But this scale change in the measure of time does not affect the workings of the clock, that is, the mechanism behind the face of the clock is not affected! It is the mechanism of the clock that is analogous to the aging of a human body, not the reading on the face of the clock!

MO: Then in the twin problem, each twin, Peter and Paul, when he sees his brother in motion relative to him, must use a contracted time measure in order to express the physical laws in the reference frame of his brother. I see, then, that this interpretation of the 'contraction formulas' for the time measures does not at all imply any biological effect on one of the brothers different from the other. It means that there is no implication here that there would be an asymmetric aging effect because of the relative motion, and therefore there is no paradox — the twins would remain the same age throughout the round trip journey of either of them, relative to the other. I am glad to see that logic is restored and the theory of relativity is indeed a logically consistent scientific theory.

JACKIE: Yes, this is the idea. With this interpretation of the time and space transformations from one reference frame to any other relatively moving one, as not more than scale changes in the expressions of the physical laws, so as to maintain their form in one-to-one correspondence in all possible reference frames, there is no logical paradox encountered in the theory of relativity![2]

MO: Whew! I am sure glad of this. After all, the theory of relativity has correctly predicted all sorts of physical consequences in agreement with the empirical facts. If the logic of this theory was shown to be wrong, we would have had to search for a new physical theory to explain all of these facts anew!

MANNY: There we go again! — insisting on logic in the laws of Nature!

MO: OK, my friends, we have gone over the time for this dialogue. We'll continue next Tuesday at four o'clock in the afternoon. Let's hope that the rain will stop by then; otherwise we may have to come to the campus by boat!

Manny, Mo and Jackie stopped momentarily at the east window of the Conference Room. The view was quite different than it was about an hour earlier, when Mo was waiting for his colleagues. The bright sky, the snow-capped mountains and the bright velvet green lawns of the campus had given way to the heavy rain that rolled into the landscape. Its intrusion masked the beautiful view of the Oregon countryside, that they knew was there, even though it couldn't be seen. The outside world looked homogeneous, wet and gray at the time, quite different than Mo's earlier view. Mo wondered if there were no mountains or blue sky at that moment, because they weren't to be seen! He could not help feeling himself swerving in the direction of Jackie's realist attitude — perhaps until the next meeting!

ON QUANTUM FIELDS, UNIFIED FIELD THEORY AND THE MEANING OF SCIENCE

MANNY: What have you been thinking about since last week? Do you still believe, Jackie, that there is total order in the universe, necessarily, or have you seen the way of the majority yet on the fundamental role of uncertainty in physics?

JACKIE: I have a very strong intuitive feeling that there is total order in all that there is, even though we will never become aware of it all. Of course, we cannot, because we are only finite. Those who believe that they have learned all that there is to learn must think of themselves as omniscient! I mean, they must think of themselves as God! In any case, I still do have an unprovable faith in the existence of a total order in the world, independent of what the consensus of opinion may be and independent of the fact that we can never become aware of it all, or even beyond the smallest part of it.

MANNY: And what about you, Mo? You seemed to be swerving toward my view two weeks ago, but you favored Jackie's approach at the end of our last dialogue. Have you thought about it since then and decided on the truth! — that is, that I was right all along?

MO: I'll tell you, Manny, I think now that I agree with both of you! I like the feeling that there is total order in the universe. On the other hand, some of the empirical evidence does seem to support the idea that there is some fundamental randomness in the world, not reducible to order, whether we like this or not! We must be objective, Jackie!

JACKIE: Look, Mo, in regard to the actual data of experimentation, I don't see that there is any more reason to believe that there is irreducible uncertainty than there is to believe in total order. Either way, we must proceed on faith! Whoever says that he doesn't base any of his knowledge on faith is totally wrong! We must have some faith that a particular path of inquiry is worth the time and effort that must be spent on it. It is a faith that this way has a reasonable chance of success, though we can never say that we know the answer for certain!

I think that a good example of this is Einstein's own attitude in science. He had faith that his theory of relativity was true to Nature because of his belief that its underlying assertion — the principle of relativity — is so simple, conceptually; for this reason, Einstein believed this theory is most likely a truth of the universe. So he decided to spend most of his life in working out its implications.[1] One of these implications was a denial of the claim that there is uncertainty, or fundamental probability, in the laws of the universe. That is to say, the implicit order in Einstein's view of the physical universe automatically ruled out the quantum theory, because of its rejection of causality.

MANNY: What about relativistic quantum field theory, Jackie? This unifies the quantum theory with the theory of relativity, doesn't it?

JACKIE: No, it doesn't really do that, Manny. There is some acknowledgment in the physics community that there is something right in the requirements of the theory of special relativity. So the attempt is made to express the quantum theory of particles and radiation so that it is the same in all possible inertial frames of reference, according to the requirement of special relativity. However, when this was first done, the resulting mathematical expression, called 'quantum field theory,' was found to have a major flaw!

MO: What do you mean, 'major flaw,' Jackie? Quantum field theory has been very successful in predicting all sorts of experimental facts — especially the part of quantum field theory dealing with electrodynamics, called 'quantum electrodynamics.' This theory has successfully predicted effects with awesome accuracy, in agreement with experimentation. Doesn't this convince you of the truth of the approach of the Copenhagen School? — an approach that easily incorporates the idea that randomness is basic in the laws of Nature!

MANNY: In fact, I would go so far as to say that quantum electrodynamics is the most successful theory that we have ever discovered in physics, in all of our history!

JACKIE: Don't get so excited, my friends! I am aware of all of this great agreement. But there is still a major flaw in quantum field theory! It is the same trouble that was emphasized by one of the founders of quantum mechanics and quantum field theory, Paul Dirac. He originally saw that when the form of nonrelativistic quantum mechanics is extended to the relativistic domain, including the radiation that is emitted and absorbed by matter, the new equations of relativistic quantum field theory have no solutions!

What happened, Manny, as you know, is the following: When the mathematical formalism of nonrelativistic quantum mechanics was properly extended, at first by Paul Dirac himself, so as to incorporate the description of matter and radiation in a way that is also compatible with the symmetry requirements of the theory of special relativity, as well as the rules of quantum mechanics, it was found that all of the solutions of this formalism, called 'quantum field theory,' were infinite. Thus there were no finite solutions at all that could predict the actual physical properties of the micromatter that we study — the electrons, protons, atoms and molecules. But, don't you see? It is the set of solutions of these laws of Nature that are supposed to represent our only contact with the real world! If there are no solutions of the alleged laws of Nature, then they are not true laws of Nature — and this particular theory fails! This is a plain and simple conclusion.[2]

MO: But what about the super agreement with the experimental facts, Jackie, especially in quantum electrodynamics? Doesn't this show that this is a good scientific theory?

JACKIE: This seems to be the same sort of question we had in our discussion of the twin paradox, last week, Mo. What I stressed then, that I should like to reiterate, is this from the philosophy of science: Empirical confirmation is a necessary condition for the acceptability of a scientific truth, but it is not sufficient! The alleged theory must also be logically consistent, imply reproducible experimental consequences, imply unique consequences for a unique physical situation, and so on.

MANNY: I disagree with you, Jackie, that a scientific theory has to be logically consistent. I believe that it is a faithful theory in science so long as it successfully predicts the experimental facts!

MO: Perhaps you each have different definitions of what you mean by the term, 'science.' Surely, there can be no arguing about a concept until those who argue agree on the meaning of the concept that they are supposed to be arguing about!

MANNY: I suppose that you are right, Mo. Jackie and I seem to disagree on the definition of the word, 'science.' This disagreement, as I see it, is in regard to the extent that a scientific investigation can take us in understanding Nature. It is my contention that science is concerned with formulating the most compact expression for the facts of Nature, and nothing more! It is Jackie's view, on the other hand, that these most compact expressions for the data of the natural world are only a step toward an underlying explanation for these data. I deny the meaningfulness of this claim of the existence of an 'underlying explanation.' To me, it is pure religion, it is not science, *per se*!

MO: This is the difference between the empiricist/positivist and the realist views of science, isn't it?[e]

JACKIE: Yes, Mo, it is the main difference in our philosophic approaches to science. It was also the main difference between Bohr and Einstein, in their historic debates on 20th century physics. It has yet to be resolved in the scientific community, to everyone's satisfaction.[3]

MO: I think that most of our colleagues in the physics profession agree with Manny's view, which is that of Bohr and the Copenhagen School. Exactly what do you gain with your realist view, Jackie? I mean, everyone seems to be working happily in theoretical and experimental physics, getting new results every day, holding international conferences on the latest results, that always seem to be in agreement with what it is that is expected. Then why can't you just bend to the will of the majority and join them?

JACKIE: Sometimes, Mo, you only see in physics experimentation what you want to see! What can be gained with the realist view is that new insights can develop that are not at all generated by the experimental facts, yet these can be new ideas that imply the existence of new sorts of experimental facts to be investigated in the laboratory. A good example that demonstrates this is our experience with the phenomenon of electricity and magnetism in 18th and 19th century physics.

MANNY: These discoveries of the basic relationships of electricity and magnetism, in the 18th and 19th centuries, were purely experimental experiences, Jackie. Recall that experiments were done by Coulomb, Ampère, Oersted, Faraday, Franklin, Henry, and so on, and their findings were then summarized in the equations discovered by Maxwell. The theory of electromagnetism was thus expressed most economically in the form of these field equations, irrespective of Faraday's mystical ideas about the primacy of the field to represent matter most fundamentally! It is similar to Kepler's mysticism, two centuries earlier. It didn't detract from Kepler's actual empirical findings in astronomy, just as Faraday's mysticism did not detract from the importance of his experimental findings about electricity, magnetism and optics in the 19th century. Thus we can certainly ignore the mystical ideas of great scientists, such as Kepler, Faraday and Einstein, while acknowledging their actual scientific contributions, regarding the empirical facts.[4]

JACKIE: I couldn't disagree with you more, Manny. Your word, 'mystical' — the m-word — is the metaphysical basis of the physical theory. It is essential to give meaning to the theory in the first place. What we call 'metaphysical' is just as essential to the understanding we are trying to gain, as scientists, as is the 'physical' part.

In regard to your claim that theory is not more than an economic expression of already discovered experimental facts, what about the case of the 'displacement current' term in Maxwell's equations? There was no empirical reason for this term when it was inserted by Maxwell! He added it because of the symmetry it added to the equations. The point is that after this term was added to Maxwell's field equations, they then predicted the propagation of electromagnetic radiation, including an explanation for all optical phenomena in terms of electromagnetism (already studied), but also the prediction of radio waves, that were not yet discovered. Hertz then looked for the radio phenomenon and indeed he found it!

A few decades later, Einstein found that Maxwell's equations are 'covariant' — that is, that they have a form that is independent of whatever spacetime reference frame they may be expressed in. This led him to the principle of relativity, as the basis of the theory of relativity. My point is that if the displacement current had not been inserted, theoretically, before Maxwell had any experimental justification for it, the covariance of these equations would not have been seen. The discovery of special relativity theory by Einstein may then have been delayed for an indefinite period of time! — not to mention the experimental discoveries, such as the discovery of radio waves by Hertz.

MO: I believe, Jackie, that the hint of relativity theory in the laws of electromagnetism was already implicit in the experimental investigations of Oersted and Faraday. Oersted discovered that a magnetic force is created by an electrical current. Faraday then theorized that this meant that a magnetic field of force is nothing more than an electrical field of force in motion — the equivalent of an electrical current. He then logically concluded that, since motion is purely subjective, it would be equally true to say that an electrical field of force is nothing more than a magnetic field of force in motion. That is, if A moves relative to B then it is equally true to say that B moves relative to A, without in any way altering the laws of physics! We talked about this earlier in regard to the twin paradox — that the purely subjective nature of motion was originally seen by Galileo, in his analysis of the heavens.

JACKIE: The relation of motion to electricity and magnetism was studied originally by Faraday, when he theorized the fact of magnetic induction. This experimental discovery, of course, led to the invention of the dynamo — the way of generating electrical power that so importantly contributed to the Industrial Revolution of the 19th century. But Faraday's primary intention was to discover objective truth! Even though it led to a technological advance, this was only a by-product of his fundamental discovery in science.

There was something else that came out of Faraday's fundamental investigations of electricity and magnetism. It was the birth of the concept of the unified field theory. What Faraday saw was the following: If magnetism is not more than electricity in motion, then since motion is only a subjective part of the description of electrical phenomena, one must conclude that electricity is not more than magnetism in motion, thus there is no objective electrical field by itself or an objective magnetic field by itself. Rather, the logical implication is that they must each be manifestations of a unified electromagnetic field of force, appearing as one sort of force or the other under correspondingly different sorts of conditions of observation. This was the initiation of the concept of the unified field theory. It is the approach that Einstein continued to pursue about a hundred years later as it was also implied by the logical structure of the theory of relativity, in its full form.

MANNY: Look, Jackie, modern elementary particle theory, which is based on the ideas of the Copenhagen School and its positivistic philosophy, is also pursuing the idea of a unified field theory — everyone believes it is making progress that would be to the satisfaction of Einstein's hopes! Yet, it does not at all rely on the realistic approach of Einstein! What I am saying, Jackie, is that the unified field concept is not necessarily a consequence of a realist mysticism! One still has this concept to guide physics research, without the unnecessary baggage of the alleged reality that is claimed to underlie what we see!

JACKIE: I believe that you are confused, Manny, about the meaning of a unified field theory in Einstein vis-à-vis its meaning in the modern day elementary particle physics research. They are entirely different concepts, even though they are called by the same name! Mo, you were so right in your comment that before there can be an argument on the validity of a concept, there must be agreement on what the concept is defined to be! The unified field theory concept is one thing to Faraday and Einstein, it is something entirely different to the present day elementary particle physicists. I do not believe that either would accept the validity of the other, once it is defined for them, even though they call the concept by the same name!

MO: But, Jackie, I am continually reading in the popular press, as well as in the scientific literature, that the present day theoretical schemes in elementary particle physics, such as GUT (Grand Unified Theory) is going to answer Einstein's lifelong dream of finding a unified field theory!

JACKIE: Nothing could be further from the truth, Mo. The elementary particle approach to unification is in the context of a quantum field theory. It is not even a field theory in the first place. As we've mentioned before, it is a scattering theory of discrete particles, expressed in the form of the probability calculus of quantum mechanics. The unification intended here is in the sense of adding new parameters and degrees of freedom to the form of quantum electrodynamics, in inductive fashion, rather than deriving the unification from scratch, deductively, as the principle of relativity led to the unification of electricity and magnetism. Further, the unified field theory of Einstein and Faraday is based on the continuous, nonsingular field concept, where there are not, in principle, singular particles, only a single matter field that manifests itself as one type of force or the other, depending on the conditions under which matter is observed. This is an entirely different concept than the particle view of unification, in the context of the quantum theory.

MO: I think that I see now what you mean, Jackie, regarding a basic difference in concept of what Faraday meant by a unified field theory, compared with the meaning in elementary particle physics today. Let me see if I can express this in my own words.

If an observer should be at rest with respect to an electrically charged pith ball, she would detect only an electrical force, using another electrically charged test body. But if this observer should be moving with respect to the pith ball, without in any way touching it, she would also detect the effect of a magnetic field, where she is. For example, carried compass needles would align themselves in a special way when the observer is moving with respect to the electrical pith ball. Thus, the electrical and the magnetic fields of force were both there all the time! She just didn't see them both until she altered her mode of observation. It would be like looking into a large room through a small keyhole. All that she would see would be the wall opposite the keyhole, even though there is a lot more to the room to be seen, if only the observer would change her way of looking into it. Perhaps she could arrange to have a periscope inserted through the keyhole. She would then certainly see more than she did before!

JACKIE: Faraday's idea, as well as Einstein's, was then that all of the forces of Nature are implicit in a single field of force, that manifests itself as one type of force or another, or some combination of forces of particular types, when there are particular, corresponding physical conditions set up that would optimize observing their effects. All the while, however, all other possible types of physical forces are there; it is just that they are not being observed at the time!

I know from his writings that Faraday believed that the gravitational force was also unified with electromagnetism. In fact, he tried to do some experimentation that would prove this. He was, of course, unsuccessful in this because, as we now know, the magnitudes of the electromagnetic and gravitational forces are so different, under the conditions of his experimentation. But he, nevertheless, believed that there was such a unification in a general field of force. If he were here today, he would say that the short range forces of the nuclear domain, that we have discovered in this century, that is, the strong and the weak forces, are also entailed in the general field of force that he sought, and that Einstein continued to seek in our own time. Of course, Einstein knew about the short range forces. He felt that if he could succeed in unifying the electromagnetic and the gravitational field of force, the other phenomena would emerge, as well as the form of quantum mechanics, appearing as some sort of linear approximation for a theory of matter expressed in general relativity. However, he was, unfortunately, not successful in his lifetime in unifying the electromagnetic and the gravitational force manifestations.

One of his great successes was to re-express a theory of gravity in the form of a continuous field theory — his initial expression of the theory of general relativity. This theory succeeded in replacing Newton's theory of universal gravitation — replacing the concepts of atomism and action-at-a-distance with the continuous field concept and the concept of forces propagating at a finite speed between interacting components of a continuous matter field. But he felt that this success was only a first step toward the unification of gravity with the other forces of Nature, in the context of general relativity as a theory of matter.

So you see, Manny and Mo, the idea of a unified field theory in Faraday and Einstein was that there is a general field of force that is expressed as a solution of the most general form of the field equations — the laws of Nature — that has implicit in it all possible types of force field, even the types of forces that may not have yet been discovered in physics experimentation. In fact, if the general form of the field could be determined, it might predict new sorts of forces that we haven't yet dreamed of, for example in dimensions larger than we have ever seen in astronomy of the universe as a whole, or in dimensions much smaller than we have yet experienced in elementary particle physics.

MO: On the other hand, Jackie, the concept of unified forces, according to modern day elementary particle physics, is different than this. Isn't that right, Manny?

MANNY: Yes, it entails a different field corresponding to each type of particle in the system. These are the 'quarks,' the leptons (electron, muon and neutrino, and their antiparticles), the W and Z boson fields that are supposed to be involved in the weak-electromagnetic interactions, and so on. The idea here is to put all of these different sorts of particle fields together in a soup, that incorporates the different kinds of forces in a total type of description for all of it. Eventually, the hope is to also incorporate the gravitational force with a particle field of a different sort, called the 'graviton.' This would be a quantum of the Einstein field, that already correctly describes gravity on the macroscopic scale. However, such a unification with gravity, in the context of the quantum theory, has not yet been demonstrated. It is the object of the so-called GUT theory (Grand Unified Theory), to derive this sort of unification.

I get the feeling, Jackie, that you believe that this sort of unification is inferior to the unified field theory that Einstein and Faraday had in mind. Do you really feel this way, and if so, why do you think that one type of unified theory is superior to the other, if they each would give equally successful agreements with the data?

JACKIE: I'm certainly not saying at this point in time that one of these approaches to a unified theory is superior to the other on experimental grounds. What I am saying is that the concept of a unified field theory in Einstein is entirely different than the unified theory according to the present day particle theory. Each entails different concepts of science and different types of logic (one based on an inductive method toward the discovery of scientific truth and the other based on deductive logic). They also each entail widely different sorts of mathematical expressions. In principle, then, the different types of mathematical language implies that there would be different sorts of physical predictions coming from each approach, for the same physical situation. Thus there could be experimental tests that would differentiate one of these approaches as superior to the other, if the experimental tests would agree with one theory and not the other.

MO: All right, my friends, the time has come, unfortunately, to leave the cosmos and return to our families. We will meet next week, same time and same place.

Manny, Mo and Jackie took their usual pause at the Conference Room windows, each thinking silently about what might possibly lie beyond the scenes they perceived — the silvery, rushing Willamette, reflecting the rays of the half moon that evening, and the deep blue night sky revealing the contrasting shadows of the Cascade Mountain range in the east. Their pause had developed into a silent, meditative ritual, when each of them pondered privately about how their questions in physics may truly relate to the real world out there! One conclusion that they all reached was that the world is indeed beautiful, and that a valid scientific explanation should reveal it!

WEEK 5

POSITIVISM, REALISM AND HOLISM IN RELATIVITY AND QUANTUM PHYSICS

Manny, Mo and Jackie arrived together with excitement in their steps. It was precisely four o'clock in the afternoon this late fall day. They were gazing out of the western window of the Conference Room. Only the upper rim of the sun could be seen, softly falling below the horizon. There was a bright orange border along the top of the rushing Willamette River. The blending of Nature's colors was fabulous — the deep blue of the river, with its white caps, the orange of the sun, fading into the dark night sky, the dark green of the campus lawns. Were they separate things? Or were they all manifestations of a single glorious entity?

MO: We certainly covered a lot of ground last week, my friends — the question of what science is really all about, the difference between the meaning of a unified field theory in Einstein and in the modern elementary particle theories. But I don't know if anyone changed their minds on anything! Manny, you seem as firm as always on the empiricist point of view, where there is a basic role in the laws of Nature for the concept of uncertainty. And you, Jackie, seem as convinced as ever about the realist view, assuming that there is a real world, independent of any observers, and that its laws are based on a total order. You probably think this because it is a beautiful concept to you! Am I right about this?

JACKIE: It is not beauty alone, Mo. There is also a strong practical element in this scientific point of view. It has to do with the mathematical languages that must be used to describe a closed system and an open system. If these languages are different because of their different approaches to physics, then they would generally make different predictions in regard to some of the physical experimentation. Thus, one should be able to distinguish which of these theories is true and which is false, as a truth of Nature, or if, indeed, they may both be false!

MANNY: I think that this is an illusion, Jackie. Suppose that we do have different mathematical languages, one for your closed system approach to the universe and one for the open system approach that I support. Suppose that the experimental facts support your formalism and not mine. What would this mean to me? Does it mean that my entire physical approach should be abandoned? No, I don't think so. All that I would have to do to come back in line with the observed facts of Nature would be to add some new mathematical language with some new parameters to adjust to the new data. Then, we'd agree once again! There would be no basic reason to accept your philosophical approach of a closed system over my view of an open system!

JACKIE: Perhaps it is correct, Manny, that you can always 'cook up a curve' to fit any new experimental data. You would then have described the data. But this does not mean that you would have explained it! There is a famous comment that if you give me three parameters I will be able to draw an elephant for you. If you give me one more parameter, I can make its tail wiggle! But would I have explained an elephant?

MO: I see what you mean, Jackie. It may be compared with this example, Manny. I read that a computer program has been written that can duplicate the shape of a beautiful woman, from any angle, and in motion! It comes out as a hologram on a TV screen. She moves in very provocative ways, throws kisses, seems to be communicating with the observer just as a real live woman would do. Now, a human female may be very well described in this hologram, but only superficially. There is so much about a woman that is not in this description, as anybody must know. The point is that there is a partial description of a beautiful thing of Nature here, but there is no explanation in terms of underlying laws of Nature! — that is, the law of a human being.

MANNY: You are certainly right, Mo. All that the computer programmer did here was to picture the outside of a woman and some of her motions. She still has an inside! — a whole bunch of machinery to keep her body going, to conceive children, to think analytically, to love. . . These things are not simulated in the computer program. But if we could simulate all of that machinery that is a woman, which, I am sure, is probably not in principle possible, then we would have explained 'woman.'

JACKIE: What you mean, Manny, is that you would have described a woman more completely than the hologram did. But you would still not have 'explained' her! I can personally attest to that!

MO: What is the difference, Jackie, between a description, if it is total, and an explanation?

JACKIE: An explanation is based on underlying principles that, in turn, lead to the description, as particulars of a universal — to use the language of logic. These would follow deductively and conclusively, given the assumption about the truth of the starting principles, i.e. the principles of Nature that are being investigated.

An example that I have been thinking about is the concept of holism, as a fundamental principle of the universe. Look over there, toward the Chemistry Building. There is a large puddle of water left over from yesterday's rain. If you look closely at the puddle, you'll see that when bits of matter from the tree fall on it, ripples are produced. Well, what about an individual ripple of this puddle? Can we describe it completely, as a thing-in-itself? Could we remove it from the puddle and measure its properties — weight, size, and so on? Why does the ripple move the way that it does? What is happening when it collides with other ripples of the same puddle?

MANNY: Of course, as physicists, we know that the actual motion of the ripple is governed by the laws of hydrodynamics. This is a continuum theory that entails the entire puddle, all at once. Thus, there is no meaning to the ripple, as a thing-in-itself. To understand the ripple, we must understand the entire puddle, and look at the ripple as one of its modes of behavior. The ripple is of the pond, it is not a thing in the pond!

JACKIE: Exactly! This is an example of a closed system, Manny. Then why do you exclude the possibility that the entire universe may be of this nature? This is the idea that there is only one universe, without separable parts. What appear as separate things are really manifestations of the whole entity that is the closed universe.

MO: This is the idea of holism, isn't it, Jackie?[f] I've read about this philosophy in the Asiatic thinking — in Brahmism, Buddhism, Taoism. But I find it hard to believe that this point of view applies to the real, material world! Look, Jackie, this is a view that implies that, in reality, there are no individual entities at all! Well, what about us? We are not like the ripples of the puddle! We are individual things, with free wills. We are not really tied to the rest of the universe, in a fundamental way! We make decisions every day, such as a choice to go to a play, that don't depend on our distance from our neighboring galaxy, Andromeda!

JACKIE: Your doubt about the reality of holism reminds me of a conversation I was told about by our colleague, Scott Blumfield, with his six-year-old daughter, Suzie. One morning at breakfast, Suzie asked, "Daddy, where was I before I was born?" Scott replied as any hard-nosed scientist would, full of confidence in his fatherly voice, "you were nowhere before you were born!" Then Suzie replied with just as much confidence, "I don't believe that, Daddy! I can believe that you don't know where I was before I was born, but I must have been somewhere!"

Scott then took a clean napkin from the table and showed it to Suzie, asking, "What do you see on this napkin?" Suzie replied, "Nothing." Scott then took his pen and drew a circle on the napkin. He then asked Suzie, "Now what do you see on the napkin?" Suzie replied, "A circle." With the confidence of a sportsman who just landed a very big fish, Scott asked, "Well, where was the circle before I drew it on the napkin?" With the same confidence, Suzie replied without hesitation, "It was in the pen!"

Scott told me that Suzie made him feel very frustrated because he was not able to explain a very simple fact to her! He said that he knew that she was very intelligent, but he could not understand why she seemed so stupid at times!

I replied that perhaps Suzie was more correct about this than he was. For the next succession of questions and replies might have been: "Where was the circle before it got into the pen?"; reply: "It was in Scott's arm and fingers"; question: "Where was the circle before it got into Scott's arm and fingers?"; reply: "It was in his mind, that is, he had the idea of a circle in his thinking process"; question: "Where was the circle before Scott had the idea of it in his mind?"; reply, "It was somewhere in his environment. He may have been impressed by the shape of the full moon one evening, when he was young!" We can carry this on, *ad infinitum*, until the entire universe is used up in attempting to answer the original question about the explanation for the circle on the napkin!

MO: I get it Jackie. What you are saying is that holism is the idea that there are really not any independent things, apart from all that there is — the universe. It is an illusion to think that there are separate parts.

This is a fascinating idea, Jackie, but, as a physicist, it is hard to believe! We are used to studying the nature of separate things, and our egos tell us that we are apart from the rest. Our egos won't let us forget this! Just how can you expect me to believe that I am an inseparable component of a continuum? Like a ripple of a puddle? Nevertheless, Jackie, I must acknowledge that my ego could be wrong — bizarre as it may seem to be!

MANNY: I believe that there is truth in this idea of holism. It is most obvious in quantum mechanics, according to the interpretation of the Copenhagen School.[1] I believe that this is a truth of Nature, irrespective of what our egos want us to believe!

MO: I am surprised to hear this, Manny. I thought that the Copenhagen School teaches us that all that is meaningful in science are the measurements, and nothing else. Isn't this what you were trying to convince us of the other week?

MANNY: What I mean about holism, Mo, is that quantum mechanics teaches us that when we are talking about the microscopic elements of matter — the electrons, protons, etc. — it is not meaningful to separate the measurer (necessarily a laboratory sized entity) from the measured, in fundamental terms. This is a holistic theory because the language of the laws of micromatter do not relate to the micromatter by itself; rather, these laws (quantum mechanics) relate to the responses of a macroapparatus to this matter, and nothing else.

JACKIE: I don't believe, Manny, that the inseparability of the measuring apparatus and the measured matter, in quantum mechanics, is really an example of holism. Rather, it is a clear example of positivism! Indeed, if there were really nonseparability between the measurer and the measured, there would be no fundamental difference in the variables that describe each of these. That is to say, macrovariables and microvariables would be the same sorts of variables. This is not the case in quantum mechanics!

MANNY: Yes, Jackie, according to Bohr's physical approach, which has been verified in thousands of experiments, the measuring apparatus must necessarily be represented with classical variables while the micromatter must necessarily be represented with quantum variables — the solutions of the quantum mechanical equations. The positivistic element comes into the explanation of this asymmetry in the description of a measurement. What we say is that, as 'observers,' we are classical entities. Thus, the readings of our instruments, when they are measuring the properties of micromatter, must be expressed in classical terms.

On the other hand, the variables that describe the micromatter are the probability functions of quantum mechanics. They are probability-related because there can be no reciprocity between the measurer and the measured, in dynamical terms. What I mean to say is that, according to Bohr's view, there is no actual explicit force that is the coupling between the macroapparatus and the micromatter that is measured. If there were, the solutions of the equations for this force would lead to an exact prediction of the outcome of an experimental measurement on the properties of micromatter. However, the theory implies that there are no such exact predictions — as shown, for example, in the Heisenberg uncertainty relations.

Since there is alleged to be no such exact dynamical coupling between the measurer and the measured, the solutions of the quantum mechanical equations relate only to the probabilities that the micromatter will be in one state of motion or another, before and after the measurement is carried out. This makes sense, doesn't it?

MO: I think that we're getting too technical now, for the level of these dialogues. Let's just stick to elementary concepts. It seems to me, Manny, that you're still talking about an open system of 'things.' This description is non-classical only in the sense that there is claimed to be no precision in the way that these things are described, as there would be, for example, in the Newtonian description of a many-particle system. All that can be said about these things — components of the open system — must be couched in the language of probabilities. This, in turn, fits in very well with the idea of fundamental, irreducible, probability in the laws of micromatter. But I certainly don't see, Manny, that your quantum mechanical approach to elementary matter is holistic! I agree with Jackie, that this theory should be considered as purely positivistic! But is there any theory in physics that is truly holistic?

JACKIE: The theory of general relativity is a holistic approach, when it is considered as a general theory of matter. As I've mentioned to you both, earlier in our dialogues, this is a truly continuum field theory where, fundamentally, there are no singular particles of matter, i.e. 'parts' that make up the whole. At its most fundamental level, the theory of general relativity implies the existence of a continuous matter field that unifies all possible interactions between matter. The matter that interacts here are not discrete bits, as in the elementary particle theories (Democritean, classical or quantum mechanical)! Rather, the matter components are the discrete modes of a continuous whole — like the distinguishable ripples of the single puddle of water! Such a theory of matter, according to general relativity, is truly holistic.[2]

MANNY: I still don't see, Jackie, what is the difference between your holism and mine, in practical terms. After all, the data of elementary particle physics comes from measurements carried out by macroapparatuses on bits of micromatter. The rest of your story is only your vivid imagination on what you think it is that is underlying the data!

JACKIE: This is not really the case, Manny. As I mentioned a few minutes ago, the mathematical expressions of these two views is generally different. Thus, while some of the predictions of both theories may be the same, there are other predictions that could be crucially different. It is then possible to test which of these theories or the other would be more true to Nature!

MO: Can you tell us, Jackie, a specific example of such a difference in predictions?

JACKIE: There are many examples. One that comes to mind concerns the prediction of periodic motion by linear theories, such as Newton's classical theory of planetary motion, versus the prediction by Einstein's nonlinear expression of general relativity of a nonperiodic motion of a planet of the solar system.[g]

The general solution of a linear theory is a superposition of different solutions of the same equations. Thus, if the classical theory predicts a solution with a specific frequency, that is, the prediction that the planet will always return to an equivalent place in its orbit in equal times, then there will be any number of other solutions with other frequencies — there will be the fundamental frequency solution, another solution with double this frequency, a third solution with three times this frequency, etc. The general solution, then, for the linear theory, will be a superposition of all of these solutions. The idea that I wish to stress is that such a sum of periodic solutions is still a periodic solution. It still follows from the linear theory that a planet will return to an equivalent place in its orbit relative to the sun, in equal times, in each cycle.

On the other hand, a nonlinear theory of motion, such as Einstein's general relativity, predicts a nonperiodic motion of a planet, relative to the sun's position. That is, it would take more (or less) time each cycle (year) for the planet to return to an equivalent place in the orbit, relative to the sun. This effect was indeed observed in the middle of the nineteenth century, in the case of Mercury's orbit, even after the effects of nonperiodic motion due to the perturbations from the other planets of the solar system were taken into account, this effect was not explained until Einstein's theory of general relativity was applied to it. It was the nonlinear features of this theory that gave the correct result of nonperiodic motion, qualitatively and quantitatively.

Thus, the periodic motion versus nonperiodic motion predicted by the classical versus the general relativity theories provided a unique test of the truth of an open system theory (Newton's) versus a closed system theory (Einstein's).

MANNY: It appears that I should concede to you, Jackie, on this. However, I'm still not convinced that it would be impossible to use some sort of perturbation on the linear system that could reproduce the mathematical result predicted correctly by the nonlinear mathematical equation of general relativity.

MO: But Manny, that would only be curve-fitting! You still wouldn't know where this extra term, that you cooked up to fit the data, came from! That is, you may succeed in reproducing the correct description of the planetary motion, but this wouldn't be an explanation! Jackie's way does provide an explanation. Yes, I like her idea of a closed system!

This seems right to me because it models the entire universe. Look, the universe itself is a closed system because there is no outside of it, by definition! I have an intuitive feeling that the laws of any subsystem of the universe, such as the laws of atoms and elementary particles, or the laws of the solar system or a single galaxy, mirror the laws of the universe as a whole, i.e. the laws of cosmology. Since the laws of cosmology are, necessarily, the laws of a closed system, represented with nonlinearity, doesn't it follow that the laws of any subsystem of the universe must also be expressed in terms of a closed system, in terms of nonlinear mathematics?

MANNY: Why does this follow, Mo? These are two different problems, as I see it — the laws of cosmology, for the dynamics of the entire universe, and the laws of a subsystem of the universe, such as an elementary particle or a human being. Further, if the universe really is a closed system, then a human being or an elementary particle could not be dynamically separate from it! This seems to imply that to understand the human being or the elementary particle it would first be necessary to understand the entire universe, holistically! — thereby giving up any hope of understanding anything at all!

JACKIE: I don't agree with that, Manny, I believe that one may still investigate the closed system by successive approximations. If we assume, for example, that the sun has little effect on the weight of an electron studied in the laboratory, here on Earth, we may still treat the electron in terms of its approximate closed system of surroundings, that excludes the influence of the sun (and the rest of the universe) without losing too much accuracy. The immediate surroundings of the electron would be the particle-antiparticle pairs and radiation in its vicinity. These surroundings, together with the electron, may then be considered, as a first approximation, to be the closed system that is responsible for the observed properties of the electron. It would be a closed subsystem of the entire universe. As a next approximation, one may enlarge the size of the domain of the electron's neighborhood.

If such a procedure leads to predictions that agree with the empirical facts about electrons, in better agreement with the data than the other theories, then we would be justified in saying that it was a reasonable approximation to take the electron with a relatively small environment to replace the entire universe, in determining its physical properties. Such a procedure allows us to study science, within the model of a closed system.

MO: I see what you're saying, Jackie. At least this leads to a scientific approach that is different than the linear approach of the open system model of the quantum theory, and may be compared with it regarding the experimental facts.

MANNY: I concede on your point, Mo. It is better to have competition in scientific explanations than to offer only a single point of view. This would quickly turn into religion. Of course, religious ideas have their place in our domain of thinking, but they shouldn't be confused with scientific thinking. For one thing, scientific truth is contingent, and thus refutable. Religious truth, on the other hand, is based on faith, and is thus irrefutable. Without the approach of contingent truth and its refutability, as well as verifiability of scientific truths, we would not be able to make any progress in our search for further comprehension of the universe. This is a point that was correctly stressed by the contemporary philosopher, Karl Popper.

MO: I believe that we all agree on this, Manny. Perhaps this is a good stopping point for this dialogue — where we are actually agreeing on something, together!

The days were getting shorter in western Oregon. It was just after six o'clock in the evening and the night had come. There was an almost full moon, sending down its silver rays to the landscape of the campus of Leber College. As Manny, Mo and Jackie gazed upon the craters of the moon, they felt that there was indeed much more to the world than was revealed at first glance. They became aware, at that moment, that their comprehension was a little bit more than it was the week before. But it remained a puzzlement to them, as Einstein once remarked, that infinitesimal human beings were able to comprehend anything at all about this vast universe of theirs!

ON COSMOLOGY: RELATIVITY VERSUS QUANTUM PHYSICS

The view was awe-inspiring to Manny, Mo and Jackie, as they stood, spellbound, at the eastern window of the Conference Room. A thin blanket of fog filled the valley between the campus of Leber College and the base of the Cascade Mountains. The sun's rays from the west could be seen casting shadows on the mountains, whose tops had much more snow than they did the week before. It was two minutes after four o'clock in the afternoon, this late fall day. Before commencing their dialogue, Manny, Mo and Jackie were deep in meditation on the beauty of Nature before them.

MO: Where do we go this time, friends? Will it be into the depths of the inner sanctum of elementary particle physics, or do we go way up to into the cosmos, 42 powers of ten greater in dimension than elementary matter. Perhaps we should move outward this time. Maybe we will be able to get a perspective on ourselves, the specks of matter that we are in this glorious, infinite universe!

MANNY: I always knew that you were a frustrated poet, Mo. Just last week at a university-wide faculty meeting, I was trying to convince Jim Fielding, of the English Department, to instigate a new course called 'Poetry for Physicists.' There are courses at many universities called, 'Physics for Poets.' I think that it is just as important to teach physicists a little bit of poetry, to expose them to this side of the Arts, as it is to teach the poets a little bit of physics! Maybe we can help to dissolve C. P. Snow's barrier between the sciences and the humanities in this way![1] In any case, I am talking about the future generations, not ours!

Personally, I don't see the problem of the universe in such poetic terms, beautiful as it is. The universe is a collection of matter — galaxies full of stars, some stars binding planets to them, exotic things like quasars, black holes, pulsars, and whatever new things the astronomers may see next year! All of these matter distributions are moving in accordance with some physical laws, just as a ball rolling down a hill follows some concrete law of matter. It is up to us, as physicists, to find out what are the most accurate descriptions of the things of the universe. If we then wish to think of these things as beautiful, that is OK, but quite beside the aim of physics, I believe.

JACKIE: Don't you see some objective beauty in all of this, Manny? The symmetry of the heavens, the spiral shapes of the galaxies, the fantastic expansion of the universe as a whole. This seems to me to imply a sort of value of beauty that is beyond our subjective feelings about it. It is an intrinsic beauty. I think that this is something of the idea of beauty that Paul Dirac sought, in his search for truth of the natural laws.[2] What I am trying to say is that the real world is beautiful, independent of our personal thoughts about it. I believe that the search for beauty can serve a heuristic function in helping us to find scientific truth. I think that Dirac also believed this idea; it certainly helped him in his discoveries of the foundations of the quantum theory, according to his own testimony about it. It also served this function for Einstein, although he referred to it as 'simplicity' rather than 'beauty.'[3]

MO: Aside from these philosophical questions, Jackie, regarding beauty and simplicity in Nature, how did it all get started. This question has always been intriguing to me, from the physical point of view. I know what the religions say; I would like to see some serious attempts to answer this question from the view of physical science.

MANNY: Most modern cosmologists agree that there was an initial 'big bang,' when the presently observed expansion of the universe started. Extrapolating backwards in time from the present, it seems to have happened between 10 and 20 billion years ago.

MO: How do we know that the universe is actually expanding, Manny? I mean, if the universe is all that there is, by definition, then what is it supposed to be expanding into?

MANNY: What we mean by an expanding universe is that the individual galaxies of the universe are continually moving apart from each other, at an ever-increasing relative speed. The California astronomer, E. Hubble, realized this when he observed that the spectra of the elements, such as the hydrogen contained in the stars, are shifting, from one galaxy to another neighboring one. What this means is that the galaxies are in motion relative to each other. The shift of the spectra is due to the Doppler effect — that we discussed earlier. Recall that what we said was that if the source of radiation of a fixed frequency is moving away from the observer of that radiation, then there would be a frequency shift toward the lower values of the observed frequency. Thus all frequencies in the visible spectrum would be shifted toward the red end of the spectrum if the source of that radiation is moving away from the observer of it. Since the spectra are seen to be red-shifted for the galaxies that are further from us than other galaxies, it means that the galaxies of the universe are moving apart from each other.

Secondly, Hubble found that the speed of recession of the galaxies from each other is proportional to their separation, in a linear fashion. That is, when any two galaxies would become twice as far apart, they would be moving twice as fast relative to each other — this is called the 'Hubble law.' It describes the expanding universe quite accurately.[4]

JACKIE: But how did this expansion get started in the first place, Manny? I can see that if we should extrapolate back in time, the distances between the galaxies get less and less, i.e. the density of the matter of the universe gets greater and greater. In a limit, we reach the maximum density, when there was instability, causing the universe to explode outwards — the so-called 'big bang.' My question is then: How did matter of the universe get into this state of maximum density and instability in the first place?

MANNY: One might say that this was the moment when God created the universe!

JACKIE: All right, my question would then be: What was God doing before He created the universe? I mean, just how did all of the matter of the universe get into that unstable state of maximum density from which to explode? This is a question in physics, Manny, not theology!

MO: I think, Jackie, that we have no place, as physicists, to ask this type of question! As physicists, we should only be concerned with the laws of matter, as we see it. Otherwise, we are asking about the laws of matter before there was any matter! Now if there was no matter then there were no laws of physics either. Perhaps, you might say, at that earlier time there was only God! Then all there would be to talk about, in reference to the universe before the creation, would be theological!

JACKIE: This would be a cop-out, Mo. I didn't ask a theological question! I asked a physical question about physical matter: How did the matter of the universe get into the state it was in at the time of the big bang? To answer a physical question with a theological answer is a non sequitur! The physical and theological domains of ideas are entirely different contexts!

MANNY: Can you do any better, Jackie? Just how would you answer the question about the way the matter got into the big bang state in the first place?

JACKIE: What do you mean, "any better?" My point is that it is simply illogical to claim to have a theological answer to a question in physics! As I've mentioned before, the truths of the contexts of science and religion are entirely different sorts of truths. That is, the truths of physics are scientific truths, contingent on Nature and therefore refutable. Religious truth, on the other hand, is irrefutable — it is based on unprovable faith.

MO: Then how would you answer the question, Jackie: How did the matter of the universe get into the maximum density, unstable state it was in at the 'big bang,' in a strictly physical context?

JACKIE: The only answer that I can see, based on my experience, would be to say that before the matter of the universe exploded outwards, at the big bang stage, it was imploding inwards. That is to say, before the matter density of the universe was decreasing, in an expansion phase, it was increasing, in a contraction phase. The contraction changed to an expansion when the predominantly attractive forces changed to predominantly repulsive forces.

MANNY: Do you mean to tell us that there are repulsive gravitational forces as well as the attractive forces? Everybody knows that gravitation is only attractive! Newton discovered this two centuries ago!

JACKIE: If this were true, there would be no expansion of the universe! We see before our eyes, telescopes and spectroscopes, that the galaxies of the universe are repelling each other, they are not attracting each other!

MANNY: This is explained, my friend, in terms of the expansion of space itself, not by a repulsive force between the matter components of the universe! This is a consequence of general relativity theory, where space itself has dynamical properties.

MO: I think you've got this wrong, Manny. We already discussed this — that according to relativity theory, space is not a thing-in-itself. Thus it is not capable of doing anything physical, like expanding. What we saw was that space and time or any other sort of parameters we may use for the representation of the laws of Nature, are not more than a language — a set of 'independent variables,' with a logic, like the syntax of ordinary verbal language, set up for the purpose of expressing the laws of the 'dependent variables' — the solutions of the laws of Nature. The logic is the algebra and geometry of spacetime. At least, this is the view that is taken by the theory of relativity, when it is fully exploited, according to Einstein's approach — when he reached the stage of general relativity.

JACKIE: I agree with you, Mo. According to Einstein's theory of relativity, space, *per se*, cannot expand by itself. It is only there to provide a language for the purpose of facilitating an expression of the laws of matter. What we see is that the matter of the universe is repelling, not attracting — in regard to the constituent galaxies.

What you said, Manny, that "everybody knows that gravitation is only attractive," is a feature of Newton's theory of universal gravitation. It is not a feature of the new theory that superseded it as a theory of gravitation — Einstein's general relativity. In Einstein's theory, there are terms, that play the role of the force exerted by the closed system on an element of that system, that are non-positive-definite, geometrical fields. This means that under some sets of physical circumstances they can be attractive and under other physical circumstances they can be repulsive. Thus, if matter is sufficiently dense and the relative speeds between the matter constituents are sufficiently great, then the repulsive forces would dominate the attractive forces — thereby causing the explosion to occur. Then, when the matter becomes sufficiently rarefied in the expansion phase, the attractive component of the overall gravitational force would dominate and the matter would then pull together, changing the explosion to an implosion — until the maximum density is reached once again, when there would be an explosion once again — and so on, *ad infinitum*.[5]

MO: This would be an oscillating universe, continuously alternating between explosion and implosion. The 'big bang' of 10 to 20 billion years ago was then only the beginning of the present cycle of the oscillating universe. When everything comes together, just before a 'big bang,' we have what some have called a 'big crunch.' But there is not a single 'big bang' and/or a single 'big crunch.' This has been going on since time immemorial. Still, Jackie, it seems to me to go against the principle of relativity — because, just as in the single 'big bang' model, there is an absolute time when the last explosion started.

JACKIE: This time that you mention, Mo, 20 billion years ago, is only a measure from our reference frame of the universe, where we sit, here on Earth in our solar system, which is an infinitesimal part of our particular galaxy, called 'Milky Way.' According to the theory of relativity, there is no absolute time measure. According to the observations of other observers who may live in different parts of the universe, in the reference frames of different galaxies, some may equally determine that the last 'big bang,' in translating their time measure to ours, was 300 billion years ago, some other observers may determine that it was only 2 billion years ago, and so on. That is, the time measure is strictly subjective, according to this theory. What it is that is truly objective, i.e. that every possible observer in the universe would agree upon, is the oscillatory behavior of the universe as a whole. This is a feature of a truly relativistic theory of cosmology.

On the other hand, the single 'big bang' cosmology claims that there was an absolute beginning of time. There was no 'before' to talk about! This is not compatible with the relativity of time measure, according to Einstein's meaning of relativity.

MANNY: If you believe this, Jackie, then when do you think it all started, if not 20 billion years ago?

JACKIE: Who knows? Your question is not one of physics. As far as we are concerned, as physicists, the oscillations of the universe have been going on indefinitely. The quantity, 20 billion years, is only a drop in the ocean!

A theological question may be: Did God create the universe? The theological answers would be: Yes or No, or I don't know — depending on one's faith in the existence of God or the atheist's lack of faith in God's existence, or the agnostic's refusal to admit the possibility of our knowing of God's existence, one way or the other.

My point, Manny, is that these are theological questions, as yours is. The question: 'What led the matter of the universe to the maximum density state, causing it to explode?' is a scientific question that the physicist is obligated to answer in the context of physics. The 'single big bang' cosmology does not do this! Relativistic cosmology does provide a possible scientific answer to the physical question.

MANNY: Most modern day cosmologists take a view different from yours, Jackie. There was a single big bang, but it is defined in terms of quantum mechanical concepts, and particularly the subject of quantum field theory. There has also been discussion of the universe as a whole in the context of the second law of thermodynamics — where entropy plays an important role. Thus there is the idea of a wave function for the universe as a whole; there is also the idea of a fundamental measure of disorder (entropy) during the evolution of the universe.[6]

MO: Where is the logic in this, Manny? There seems to me to be no meaning at all in the concept of a wave function for the whole universe. Just recall exactly what is the context of the concept 'wave function,' according to the quantum theory. It is supposed to represent a way in which a macroapparatus responds to the measurement of a physical property of micromatter! In the case of the universe as a whole, just what is it that is playing the role of macroobserver and what is it that is supposed to be the micromatter that is observed?

MANNY: Well, maybe we should change the interpretation of the wave function in terms of measurement! The interpretation is not really that crucial. All that counts is that there is a correlation between the formulas and the data — and the correlation of quantum mechanics with microscopic matter has been fabulously successful!

MO: I don't think it is that simplistic, Manny, even if you are sold on the positivistic epistemology in physics. You see, the wave function is an element of a Hilbert space — a linear function space. If we wish to describe the universe as a whole, it is a closed system that cannot be represented in terms of a linear function space, which is crucial toward its description with a probability calculus!

The formalism of quantum mechanics, in terms of a Hilbert space, is suitable for the expression of a theory of measurement. But a theory of the universe as a whole would not be representable this way, essentially because it is a closed system at the outset. This is independent of any approximations that one may or may not use in order to estimate some of the predictions of the theory. So you see, Manny, there is not complete freedom in our choice of mathematical formalisms to represent physical ideas! In particular, there can be no linear wave function for the universe as a whole.

MANNY: I mentioned this before, that I do not rely on any law of Nature to be based on the rules of logic! If the data fits the formulas, most compactly, then it is a good scientific theory. This is my epistemological view. Perhaps this attempt to understand the cosmos, in terms of a global wave function, would be more compatible with the notion of entropy, as we understand it in the second law of thermodynamics — the assertion that any complex system, that is not in equilibrium, must increase its intrinsic disorder as it approaches the state of equilibrium, until its entropy, i.e. its intrinsic disorder, reaches its maximum value, when the system has reached equilibrium.

JACKIE: I don't understand how this would help things, Manny? The entropy concept relates to the fundamental disorder of an open system and the way this disorder increases as the system approaches the state of equilibrium! But the universe as a whole is not an open system — it is closed because it is all that there is!

There is no doubt that the laws of thermodynamics are useful in science. But they are phenomenological, not fundamental! That is to say, underlying the descriptions, called 'laws of thermodynamics,' there are fundamental relationships that are the basic laws of matter.

Let's look at entropy, in particular, in terms of an example that demonstrates the second law of thermodynamics, to show you what I mean, Manny. I believe that the following was one example used by the great American theoretician, and discoverer of subtleties of statistical mechanics, Josiah Willard Gibbs. Consider a drop of dark blue ink, inserted into a beaker of clear liquid. At first, most of the ink molecules would be confined to the space of the drop in the clear liquid — there would be maximum order in the sense that we could specify with certainty where each of the ink molecules is located. However, in time the ink molecules diffuse into the clear liquid, until the entire liquid in the beaker has turned a pale blue color. At that time, when equilibrium is reached, there is maximum disorder (entropy) because we are least able to specify where any one of the ink molecules is located in the beaker. The liquid then stays a pale blue color. That is to say, the second law of thermodynamics says that when the system is not in the equilibrium state, it will proceed toward that state, ever increasing its entropy in time, until equilibrium is reached. The entropy will then maintain its maximum value, for all future times.

MANNY: Well, isn't all of this true, Jackie? Isn't this the way Nature really operates? If it is, then the second law of thermodynamics, in its domain of application, is just as good as any other law of Nature in its particular domain of application, such as Einstein's theory of general relativity, regarding the laws of gravity.

MO: This attitude, Manny, is certainly in line with your positivistic attitude that features of a real world, independent of direct observations, are excess baggage, not necessary in science.

On the other hand, Jackie has argued for practical features of the approach of realism in physics, particularly predictions that would not even be suspected to exist in the empiricist approach that you take. How would you show that there are such practical features in the ink drop example, Jackie?

JACKIE: Look, my friends, when the ink drop starts to diffuse into the clear liquid, does it do so because it is more probable for the ink molecules to be in the bigger volume of the clear liquid than in the smaller confinement of the original ink drop? Is this the fundamental reason for the diffusion process that we observe in this example?

MO: No, this is not the fundamental reason that it diffuses, Jackie. It is because there are forces exerted by the molecules of the clear host liquid on the molecules of the ink drop that create a pressure unbalance on the ink molecules, forcing them into the space of the clear liquid.

JACKIE: Of course, Mo. Then there is a predetermined, ordered law for the behavior of the ink molecules, as they interact with the molecules of the clear liquid. But, as observers of this diffusion process, we are not aware of this order, so we measure the process of the approach toward equilibrium in terms of what we don't know — defined as the entropy of the system. Still, whether or not we are aware of the implicit order of the system, it must be there!

Another example is the shuffling of a fresh deck of cards. At the start, the deck is perfectly ordered in terms of the sequences of numbers and Jack, Queen and King, and in order of suits: spades, hearts, diamonds and clubs. Then we shuffle the cards, displacing all of them in a random order. It seems then that we have created a maximum amount of disorder, if the cards were shuffled well. But this is only disorder in terms of what we don't know — it is a subjective disorder! For if we were aware of the initial conditions on the individual cards in the shuffle and the forces that were exerted, we would know with certainty the location of each card in the deck, after it had been shuffled. If we could then apply the appropriate boundary conditions and forces on each of the cards of the shuffled deck, repeating the shuffle could reverse the process and reshuffle it into the original ordered deck. But we are unable to do this because of our lack of skill! That is, it is not an intrinsic disorder (entropy), independent of the observer, in the shuffled deck of cards. But it is indeed convenient for us to talk about our lack of knowledge of the exact place of each of the cards in the shuffled deck in terms of the entropy idea. We see, then, that entropy is a subjective concept; it is not an idea that is applicable to a closed system (e.g., the universe), in which in principle there are no separate 'subject' and 'object' at the outset.

MO: I see what you're saying, Jackie, applied to the problem of cosmology. If the entire universe would be perfectly ordered at the outset, as the oscillating universe model is, according to your relativistic description, then its evolution would have nothing to do with the concept of intrinsic disorder. There is no outside and inside of the universe, as a whole. It is only that if, for convenience, we should study the universe from the point of view of parts, and focus on only one part, we would be ignoring the other part, thereby introducing a certain amount of disorder into the total picture. This may be convenient for the investigation under way, but it would not relate to the total order of the universe, as a whole, that is there, whether or not we are able to study it! What I see, then, is that the concept of entropy, and the second law of thermodynamics, are simply out of context in relation to the problem of cosmology — the science of the universe as a whole.

MANNY: This is your opinion, Mo. I don't see it this way. I think that the entropy concept is applicable because our measurements do reveal a certain amount of disorder in the universe, and the equations that represent the laws of thermodynamics apply to what we do observe. It is the same reason for saying that the quantum theory applies globally, that is, it applies to the universe as a whole. All that we have to do, to say this, is to alter our initial interpretation somewhat. We can always do this because, as I see it, interpretations are dispensable, so long as the correct mathematical description of the data remains.

MO: What are some of the quantum mechanical applications to the problem of cosmology, Manny?

MANNY: A recent theory of the universe that has been very interesting to the cosmologists is called the 'inflationary model.' It entails an attempt by the believers of the quantum theory to understand the first few microseconds of the 'big bang' from the point of view of the latest thinking in elementary particle physics, combined with the formal structure of Einstein's theory of general relativity.[7]

What is said in this approach is the following: 'Initially,' that is to say, when there was no meaning yet for the concept of 'time,' the universe was a quantum vacuum, made up of very tiny units in a ten-dimensional space, units that are called 'strings.' The ten dimensions are the ordinary four dimensions of spacetime and six more dimensions that entail a unification of the four fundamental forces — the gravitational and electromagnetic forces, that normally are long range in the ordinary four-dimensional spacetime, as well as the weak and strong (nuclear) forces, that are normally seen to be effective only at very small distances, the domain size of elementary particles.

At this 'initial' stage of the universe, then, the mutual interaction of these 'strings' was nonsingular in the sense that it did not blow up at zero mutual separation. This is because the strings, unlike the point particles of ordinary matter, are not singular where they are, because they are somewhat smeared out in the ten dimensions.

The perfect symmetry of the ten-dimensional strings that comprised the universe was then said to be 'broken' by random quantum fluctuations of this infinitesimal cosmos, whereby the string world was suddenly inflated from the domain of micromatter to that of cosmic proportions as we see it today, yielding the formations of galaxies and clusters of galaxies, and so on in the entire universe. The point is that, according to the quantum theory, there is a finite probability that such quantum fluctuations can occur. The actual beginning of the universe was then governed by a probability of a quantum fluctuation!

Thus, the initial symmetry of all forces of the string universe was broken to a lower symmetry, whereby the gravitational and electromagnetic forces became long range, extending themselves over the entire domain of the visible universe in holding everything together, and the weak and strong forces became short range, effectively extending themselves only over the dimensions of nuclear proportions — a factor of about 42 powers of 10 smaller! When this happened, the strings dissolved into gravitational radiation, in the form of 'gravitons' — the quanta of Einstein's gravitational field. Thus, the ten-dimensional space was reduced to the four-dimensional spacetime that we now experience. Such were the changes that were supposed to have taken place in the order of $1/10^{35}$ seconds!

The physical processes that were involved in this crucial 'inflation' of the universe are said to be analogous to the phase change in the state of matter that occurs at a critical temperature, such as the change from ice (at the lower temperature) to water (at the higher temperature), whereby the energy of latent heat plays the significant role of restructuring the universe, from the totally symmetric system of strings to the asymmetric structure of ordinary matter that emerged from the moment of the 'big bang' to the present. This is similar to the symmetry breaking that occurs at the critical temperature when the more symmetric crystalline structure of ice changes to the less symmetric molecular structure of water in the liquid state.

As I understand this scenario from the initiation of the big bang onward in time, that is, when 'time' became real, the assumption is made that the gravitational force between ordinary matter can only be attractive, as in the Newtonian theory of universal gravitation. It is then postulated that, initially, there was only the physical vacuum, and that out of this there appeared a new sort of matter field in the first few microseconds of the 'big bang.' What is different about this new matter is that, aside from its instability, it repels ordinary matter instead of attracting it. This is the 'Higgs field' of present-day elementary particle physics — an ingredient that is essential to the validity of the subject of quantum chromodynamics — the so-called 'generalized gauge theory.'

MO: As I see it, then, Manny, in the first few microseconds there was this Higgs field that then proceeded to decay into ordinary matter, that, in turn, was repelled by the remainder of the Higgs matter field, until the Higgs field was all used up, leaving the ordinary matter to continue in its expansion. This is the scenario, then, that is meant to explain the presently observed expansion of the universe, as well as some other astrophysical consequences. This seems to me a quite interesting cosmological theory, don't you think so, Jackie?

JACKIE: Yes, Mo, while the inflationary cosmological model does seem to be an interesting speculation about the formation of the universe, there are a few criticisms that should be taken into account. The first point is that there is no empirically or mathematically conclusive evidence for the inflationary model of the initial universe, nor for the ten-dimensional strings that were supposed to comprise it.

Secondly, there is no evidence for the existence of a Higgs field — though a great deal of work has been done in elementary particle physics experimentation to find it!

Thirdly, when the theory of general relativity alone is fully explored, it is seen that there is no need for a new matter field, such as the Higgs field, to explain the initial repulsion of matter, when the presently observed expansion of the universe started. As we have discussed earlier, the fields that play the role of 'force' in Einstein's theory are not positive-definite, that is, under some conditions these forces can be attractive while under other conditions they can be repulsive. Thus, contrary to Newton's theory of universal gravitation, where the gravitational force is only attractive, Einstein's general relativity — even as it was formulated in 1915 — predicts in principle both attractive and repulsive forces. The theory of general relativity then entails the possibility of explaining the empirical facts of cosmology — the observed expansion of the universe — in terms of a general force of gravity that has both attractive and repulsive manifestations, with one of these dominating the other under correspondingly different sorts of physical conditions.

A fourth point of criticism is that there is no evidence, to this date, that the theory of general relativity is either logically or mathematically compatible with the full form of the quantum theory. That is to say, the inflationary cosmological model of the universe is based on the assumption that there exists a consistent quantum theory of gravity. Not only has this possibility never been established, if the gravitational theory is Einstein's field theory, but there are bona fide technical reasons to believe that such a unification is indeed impossible to achieve!

MANNY: Look, Jackie, in the more detailed analysis of the inflationary universe cosmology, which I believe to be true, in spite of your criticisms, the Higgs quanta are the source of the energy-momentum tensor of Einstein's field equations, that in turn give rise to the Robertson-Walker type solution that has already been found to yield the big bang model and the correct Hubble law for the expansion. These facts are empirically true, Jackie, then what can be wrong with it?

JACKIE: Let me remind you again, Manny, that a scientific theory cannot tolerate logical or mathematical inconsistency, in spite of correct empirical agreements. That is, we all know that for a scientific theory to be true it is necessary for its predictions to match the empirical data — but this is not sufficient! For the theory must also be consistent.[h]

In regard to the Robertson-Walker solution of Einstein's equation that you refer to, let me remind you, Manny, that its derivation was based on the imposition of several new assumptions, on top of Einstein's principle of covariance,[i] that seem to be in contradiction with the principle of covariance!

One of these is the idea that leads to an absolute time measure in the problem of cosmology — a measure that is commonly called 'cosmological time.' This is a time measure that does not fuse with the relative space measures in the form of a subjective spacetime measure, as is required by the full exploitation of Einstein's principle of covariance.

MO: Look, Manny, with this view of cosmological time, all time intervals are Newtonian — in the sense that their measures may be considered with respect to an absolute beginning of the universe. This is the singular beginning of the 'big bang,' according to this cosmological model.

On the other hand, the principle of general covariance rejects this concept because all time measures, from the view of any spacetime reference frame, must be subjective, i.e. they are all relative to the reference frame from which they are taken. Thus, the absolute time of the inflationary universe cosmology is an interesting speculation that is compatible with the conceptual basis of Newtonian cosmology, but it is incompatible with a cosmology based on Einstein's theory of general relativity.

As I see your view, Manny, in the beginning there was no matter at all. There were only ten-dimensional strings. These strings then were equivalent to a curved spacetime, in conformity with Einstein's theory of general relativity. But the curvature of this spacetime, in Einstein's theory, is equivalent to energy-momentum. Because of its dynamics, the space of the initial universe had a finite quantum mechanical probability of making a transition to a mixture of ordinary matter, such as the quarks that now bind together to make up protons and neutrons, and Higgs-matter. This is because matter is equivalent to energy, according to Einstein's theory of special relativity. Once the Higgs field appeared, some of it decayed, in a short time, to ordinary matter. This created a soup containing Higgs matter, that had not yet decayed, and ordinary matter. The Higgs matter then repelled the ordinary matter, starting off the expansion of the universe, as we presently see it. In a matter of microseconds, the Higgs field totally decayed, leaving only the ordinary matter that we now see — protons, helium, and so on, that make up the galaxies and clusters of galaxies and all other material components of the universe. This is the sort of scenario that explains the expansion of the universe since the initial 'big bang.'

JACKIE: You shouldn't say that matter is equivalent to energy, Mo, or that space, *per se*, could have energy. Space, according to the theory of relativity, that you are drawing on, is not a physical entity! It is only a relative language whose sole purpose is to facilitate an objective expression of the laws of matter, that is, the laws of Nature. Matter is the substance of the universe; it has many physical properties, energy is one of them — it is the ability of matter to do work. Because the sky is blue, would you say that 'sky' is equivalent to 'blue?' The energy-mass equation of special relativity tells us that the inertial mass of a quantity of matter is 'a measure of its ability to do work,' that is, its intrinsic energy. And this is what Einstein said about the equation in his 1905 article that derived it — that mass is a measure of the energy of a quantity of matter, not that mass is equivalent to energy!

Aside from this logical error in the inflationary universe cosmology model, am I wrong, Manny, that there is no tangible evidence for a Higgs field? Was it cooked up in order to fit the data of an expanding universe?

MANNY: Oh, no, Jackie. The Higgs field is an essential component in the latest thinking about elementary particle physics. It was discovered, theoretically, long before it was applied to the problem of cosmology. There are many experiments in progress today, in high energy particle physics, that are attempting to find the Higgs particle — a quantum of the Higgs field. However, you are correct, Jackie, that there has been no evidence to this time that such a particle indeed exists.

MO: I have read, in some recent summaries of particle physics, that this same theory that involves the Higgs particle also predicts the existence of other things — that also have not been found, to this date. One of these is the magnetic monopole — like a single, isolated north pole of a magnet, separated from a single south pole. It is a crucial ingredient of this theory of elementary particles, called 'quantum chromodynamics,' that there should be an elementary particle that is a magnetic monopole.

There is also the prediction in this theory that the proton is not a stable elementary particle — essentially because it is a composite of other sorts of particles called 'quarks' that are confined to its volume, and there can be excitations of this composite that cause the proton to decay. Though the average decay time of the proton is very long, the theory predicts that, for a very large number of protons, a few decays should be observable. Have these things been seen yet in high energy physics experimentation, Manny?

MANNY: Unfortunately no. But I am quite confident that the high energy physics experiments now planned will yield the evidence for these things.

JACKIE: Don't you think that it is possible, Manny, that quantum field theory is just not one of the truths of Nature, and that the reason for our not seeing evidence for the Higgs particle, the magnetic monopole or proton decay is simply that they don't exist, and further, that these things have nothing to do with the problem of cosmology?

MANNY: Your skepticism is understandable, Jackie. But I'm sure it will be seen to be ill-founded. After all, so many bright minds today in experimental and theoretical physics believe that the 'Standard Model' that predicts these things, has already been proven in large part in Nature.

MO: And what about gravitation, Manny? Does your quantum field theory rigorously incorporate Einstein's theory of general relativity? It seems to me that this would be essential for your 'inflationary cosmological model!'

MANNY: This has yet to be done! Only time will tell! But I do believe that all of this will be confirmed in time. It has to be so! It would substantiate Bohr's original view of physics, as well as the work of the majority of physicists.

JACKIE: I'll put my money on Einstein, in the long run, Manny — even though the consensus today is against it! As you say, time will tell!

MO: One thing that is clear, colleagues, is that we will never reach any complete statement about the laws of Nature, in any domain. Three cheers for controversy — it keeps us discovering! I recall reading Galileo's comment, in his Dialogues, that "There is not a single effect in Nature, not even the least that exists, that the most ingenious theorists can ever arrive at a complete understanding of it." I think that this is because the amount of understanding that would be necessary to be complete, for any phenomenon in Nature, must be infinite. But one thing that we can do, as finite human beings, is to increase our understanding, even though we can never complete it. I believe that our dialogues are a very effective means toward that end, because they heighten controversy, which, in turn, gives us the chance to see different points of view.

One of the most fascinating things about our dialogues today and last week, is the possibility that the laws of Nature, from cosmology to the laws of elementary matter, may be governed by a common set of principles. Whether, in the long run, they will be in your camp, Manny, of positivism/empiricism or along the approach of realism, according to your view, Jackie, has yet to be determined. But it is a most fascinating and beautiful study to pursue!

As usual, Mo's philosophizing at the end of the day led to at least one definite agreement between the three friends — until the next time!

MATHEMATICS TO PHYSICS TO PHILOSOPHY

It was that time of the week again for further explorations into the mysteries of the universe. As they waited to use the Conference Room, Mo thought about the conversation between Einstein's wife and an astronomer at the Mount Wilson Observatory in California. The scientist was showing Mrs. Einstein all of the elaborate equipment that was used to explore the furthest stars of the heavens and their implicit order. Mrs. Einstein replied, "You know, my husband studies the same problems of the heavens on the back of an envelope!"

Mo knew that his dialogues with Manny and Jackie would lead them to new knowledge, though from a different angle than experimental studies in laboratories. He was becoming impatient waiting for Professor Scott Blumfield to finish his seminar on the latest finding in low temperature physics. Though Mo had the room reserved for every Tuesday afternoon at four o'clock, Scott was running overtime this week.

As they closed the large oak door, Manny, Mo and Jackie saw the deluge of rain pelting the window. There was no visibility of the Oregon landscape that afternoon in any direction from the Physics Building of Leber College. As he stared at the water streaming down the eastern window, Mo was pondering about views on the problems of education in ancient Greece.

MO: I believe that Plato, in his *Republic*,[1] advised the student who wished to sharpen his mind in order to pursue truth, that he should follow a particular sequence of subjects: First, he should study mathematics to perfection — arithmetic, then plane geometry, then solid geometry. This would prepare him for an understanding of his next subject. When the student matures this far, he should then study physics — which in ancient Greece was a study of astronomy and an understanding of the motion of matter, as well as a study of the subject of acoustics, i.e. the relations of numbers to the harmonics of pure sound. Finally, with a mastery of physics, if the student had reached the age of at least 30, he was in a position to study philosophy — which Plato called 'dialectics.'

JACKIE: Yes, Mo, I believe that Plato defined dialectics as the procedure that leads to a rational account of the essential nature of each particular of the world. Thus, to comprehend anything about the real world, Plato argued that it is necessary to understand its essence first — that is, what it is that underlies outward appearances. This is the epistemological stand of realism — the idea that there is a real world, independent of the observer, that we may reach only in part by grasping strands of its essences, by rational means.

MANNY: It is interesting that the Middle Ages philosopher and theologian, Moses Maimonides, extended Plato's educational plan. In his philosophical treatise, *The Guide of the Perplexed*,[2] which was written for the serious student — that is, the student who must be perplexed! — Maimonides advised that once he had mastered philosophy, as Plato advised in his educational plan, the student was ready to comprehend some of the essences of the Biblical Scriptures. That is, Maimonides' sequence of subjects was: Mathematics first, then physics, then philosophy — taking account of the material world — and then religion — to take into account the spiritual underpinnings of the real world. That is, the subject of religion, in this theologian's view, raises questions that transcend the physical world — dealing with the supernatural causes for our existence, expressed in terms of God and our relation to Him.

As far as I am concerned, however, as a physicist, I remain perplexed! I don't really see that religion or philosophy, or for that matter, mathematics, are primary for our study of the physical universe! I don't deny the importance of religious and philosophical truths for our overall knowledge. But I don't believe that they indeed relate to the truths of science, *per se*!

MO: I agree with much of what you have said, Manny. Still, I wonder how you can say that you really know something if you are unable to express it. How can the laws of Nature be expressed in their full form without the mathematical language?

I recall reading in Galileo's Dialogues his comment that "God's Book of Nature is written in the language of mathematics." I do agree with Plato that mathematics is essential for us to know, in order to comprehend the real world.

MANNY: What I meant to say, Mo, is not that we don't need mathematics. It was only that mathematics plays a secondary role in regard to the essence of science, not a primary one. Mathematics is only a tool whose purpose is to facilitate an expression of the data of our measurements, in the neatest fashion possible. But who is to say that it would be impossible to find a different means of expression of these measurements of science some day? The data is essential for science. However, the particular language that is invented for the purpose of expressing it need not be essential since other, more efficient languages may be found later on as the field of mathematical and verbal expression are developed further.

JACKIE: I agree with you, Manny, that to simply have an awareness of the mathematical language, subtle as it may be in all of its intricacies, is not in itself to know Nature! We have seen how the history of mathematics, ever since ancient times, has revealed increasingly sophisticated logical systems. First, in ancient times, there was arithmetic — the calculus of whole numbers — and on to irrational numbers, then to geometry, and so on until the Renaissance period when calculus was invented in order to describe variable motion — and thence to a rapid development to our time, with the creations of algebraic topology, differential geometry, and so on. A great deal of this mathematical development has been useful in science. But a great deal of it has not yet been useful, nor perhaps ever will be.

Still, we must not lose sight of the fact that the subject of mathematics is a field of intellectual development in itself. It is beautiful as a logical set of systems, yet it is not science. That is, mathematics, *per se*, does not deal with the contingent laws of Nature. Its truths are necessary.

MO: I think that mathematics is just as much a part of Nature as physics is. Why do you demean this subject, Jackie — putting it on a level below science, as Manny also seems to be doing?

JACKIE: Look, Mo, we are not trying to demean mathematics when we say that it is not science! We are merely saying that it is a subject that is not in the same context as science. Mathematics seeks logical conclusions that are necessary truths — truths that cannot possibly be false, given the axiomatic basis that leads to them, deductively. On the other hand, the truths of science are not necessary truths — they are contingent on the axioms that truly characterize Nature, they are not man-made axioms! Thus, scientific truths are refutable, in addition to being verifiable. They can never be established conclusively, because we can be in error on what we think are the axioms that underlie particular physical phenomena. That is, the axioms of science transcend us — they are truths that we are trying, as scientists, to discover. The axioms of mathematics are inside of us, because they are created by us. We don't discover the truths of mathematics, we manufacture them! We do discover the truths of science, and we can never be sure that they are indeed conclusively proven. In fact, I feel certain that we can never fully comprehend the truths of any physical phenomenon conclusively, because, as Galileo implied, their extent is infinite while we are finite creatures! However, what we can do, as human beings, is to achieve progress in our understanding of physical phenomena, even though our total understanding may never be completed!

MO: Pythagoras thought that mathematical truths are reflections of the laws of the universe, just as the relations between the ratios of integers reflect the harmonics sounded by a stretched string. Euclid thought that the mathematical subject of geometry is the science of space. That is, Pythagoras and Euclid thought that mathematics, *per se*, is science.

JACKIE: We have greatly increased our mathematical knowledge since the period of ancient Greece, so much so that we are now aware that there is no exact correspondence between every mathematical theorem and every fact of Nature! That is, we are capable of creating all sorts of mathematical systems that have nothing to do with the reality of Nature, even though these systems are analytically true, in themselves. Thus, I believe that Pythagoras and Euclid, as well as many of the modern day mathematicians, are quite wrong about their belief that mathematics is science.

If space was a thing of Nature, i.e. if it had ontological status, as Euclid believed, then it should have an objective, scientific representation, independent of any of us 'knowers.' This was the thinking of Karl Gauss, the great 19th century mathematician/physicist, when he tested a result of Euclidean geometry, that the sum of the three angles of a triangle is 180 degrees. What he did was to set up posts on three mountain tops, not in a line. He viewed them with telescopes, carefully measuring the angles internal to the triangle. He found that they did indeed add up to 180 degrees, in agreement with Euclidean geometry. Of course, the sum of the angles of a triangle on the surface of the Earth should not add up to 180 degrees because it is on a curved surface while Euclidean geometry is based on plane surfaces. In any case, Gauss was performing a physical experiment to test the conclusions of Euclidean geometry, as though it were a scientific theory of Nature. He did think about the possibility of the existence of other geometries that are not Euclidean. Not long after Gauss' speculations and experimental observations, non-Euclidean geometries were indeed discovered, by Lobachevsky, Bolyai and Riemann. Still, they are just as conventional as is Euclidean geometry. Contrary to Gauss' idea about geometry being an experimentally testable science, we have now found that the non-Euclidean geometry discovered by Riemann is simply a more adequate language to express Einstein's theory of general relativity than Euclid's geometry is.

What we have discovered, with the theory of relativity, is that space is not a thing-in-itself. Rather, as we have discussed in our previous dialogues, it is not more than a language that is relative to a reference frame in which a law of matter is described, with that language. Thus, with this view, geometry plays the role of a part of the logic of the language of laws of Nature, analogous to the syntax of ordinary, verbal language.

MANNY: From what you have said, Jackie, it seems to me that what we can know falls into two distinct categories. First, there is analytical knowledge regarding the languages that we construct, such as our knowledge of the rules of arithmetic, or our knowledge of the logic of propositions, leading us to true or false conclusions about a given set of assertions. A second type of knowledge is scientific, being contingent on something that includes ourselves, but also transcends ourselves. That is, this knowledge is contingent on Nature — the universe in all of its domains — whether or not we may be aware of it at any given time. The first type of knowledge is certain. The second type of knowledge is not certain — it is continually subject to refutation.

MO: You truly surprise me, Manny! It seems that you are now agreeing that it is actually possible to refute some of your favorite theories of matter, such as the quantum theory and the elementarity of probability in Nature.

MANNY: I hesitate to agree with you on this, Mo, because the quantum theory and the calculus of probabilities in theories of matter are the data, and the data cannot be false! However, I must concede, after our discussion today, that it is remotely possible that the rules of the probability calculus of the quantum theory may have to be modified some day, to accommodate new results that may not fit the conventional formalism. Because of this refutability of the quantum theory, it may then be classified as a scientific theory, that is to say, a theory based on scientific truth rather than necessary truth.

MO: What sorts of alterations of the quantum theory might you anticipate, Manny?

MANNY: It may be necessary, some day, for example, to add some sort of nonlinear operator into the formal equations of quantum mechanics. There have been some recent results on this, such as the theory of solitons.[j] This theory indicates the possibility of generating matter waves that do not disperse when they interact with other matter. This modification may indeed answer Schrödinger's criticism of wave-particle dualism in the standard interpretation. His objection was to the idea that the matter-wave relates to a single particle of matter, such as an electron. It went as follows: If a given particle is a definite wave, then after it interacts with other matter it must disperse into many scattered waves. Then which one of the scattered waves is the original matter-wave? To derive a dispersionless wave to represent the particle may then answer his criticism.[3]

MO: Aside from the fact that there is as yet no empirical reason to believe that the solitons relate to actual electrons, that is, that the dynamics of electrons is in one-to-one correspondence with the dynamics of these dispersionless waves (originally derived from the hydrodynamics of shallow water), you are not really altering the basic structure of quantum mechanics in this example, Manny! You are still dealing with a linear function space that has the necessary structure to be compatible with the type of probability calculus required by the quantum theory, as a theory of measurement. With the addition of the nonlinear operator that you mentioned, the implicit forms of the wave solutions of the quantum theory would be altered, but you still have the same old Hilbert space description — a probability calculus to represent the fundamental laws of micromatter. What I would like to know, Manny, is this: Do you believe that it is possible that this probabilistic theory, in the form of a theory of measurement, would play no role in a foundational theory of matter?

MANNY: I do not believe that this could be true — because, as we have now realized, all of our data depend on measurements, which in turn are best represented with the formalism of quantum mechanics. This is something that we know with absolute certainty!

MO This reminds me of George Bernard Shaw's satire on modern physics: "One thing that Heisenberg knows with absolute certainty is that there is no certainty!"

JACKIE: Yes, Mo, Heisenberg's uncertainty principle is one of the pillars of the quantum theory. It is very difficult to understand for students as well as anybody else, including a brilliant playwright, like George Bernard Shaw.

Your assertion, Manny, that you know with certainty that the quantum theory is absolutely true to Nature because it represents the data, is a non sequitur! Of course it is correct to say that our data come from measurements, but why does it then follow that there cannot be other, more fundamental truths, based on a different sort of language, that lead to these same data? This would necessarily bring into our search for truth, a deeper role for the reasoning power of our minds. In this case, the comparison of the data with our theoretical predictions would only be the final stage of our search for the truth. This seems to me similar, again, to Plato's metaphor of the cave.

MO: Just how do you see this similarity, Jackie?

JACKIE: In Plato's metaphor, the slaves are tied down to the ground, only allowed to look at the wall in front of them, where they see the shadows cast by the light that diffracts from the objects or fire at the cave entrance. But then Plato considers some of the slaves breaking their chains and then turning toward the cave entrance. At first they are blinded by the glare of sunlight. However, they slowly adjust to the bright source of light. After returning to their place, facing the wall, they then know that there is something real about a source of the moving shadows on the wall. They know, at least, that they are projections of something that is real, and not things in themselves.

MO: In another of his dialogues, Plato discusses a similar metaphor that illustrates this idea of reality. It is the sensation we have of wind, vis-à-vis the existence of air. The air is there all the time, but if there were no outward variations, such as the temperature gradients that cause movements of some of the air relative to the rest, creating the wind that we do respond to, we would be totally unaware of the existence of the air — which is the underlying reality in this example.

MANNY: These may be good analogies for the concept of an underlying reality, but how conclusive are they in claiming that there is such a reality?

JACKIE: Of course, what we believe to be the reality that underlies the observables of physics can never be conclusive, in specific terms. But Plato's metaphors do give convincing arguments that there is a reality that underlies all observables, even though we can never have a complete understanding of this reality.

MANNY: You can't prove that, Jackie! Then why do you persist in believing it? It is excessive metaphysical baggage that only clutters up your mind, getting in your way of doing real scientific work!

JACKIE: It is true that we cannot prove that there is a reality, independent of our observations in science. However, I believe that it is the credo of the scientist to have faith in the existence of a reality underlying the data. After all, the time of a scientist is spent in trying to explain these data, not merely recording them! It is the reality that explains the phenomena of Nature!

If we only have the facts of experimentation in front of us, what do we really know — that is, what have we explained about Nature? Don't you see, Manny, the main object of science is to explain natural phenomena. This is what we mean when we say that we 'know' something. What you call "excessive metaphysical baggage" is, in point of fact, the main goal of science in the first place!

MO: This brings us back to the definition of science, my friends. What is it? Is it a search for a better description of Nature than we presently have, or is it a search for an explanation of Nature? To me, it is exciting to think that we, as scientists, are probing the deep mysteries that underlie the workings of the universe. But might not this be an illusion, as Manny believes? How much of what we now believe to explain some natural phenomenon was an accepted explanation 300 years ago, or 2000 years ago? If everything that we believed as true in the past periods of our history are now known to be untruths, then whatever we think of today as the truth will likely not be so in future times! Are we then fooling ourselves in thinking that we can discover any objective truth of the universe? Is science not more than a game that we play, changing the rules of the game from one period of our history to the next, but never actually discovering any of the objectively true concepts of the universe?[4]

JACKIE: No, Mo, I don't believe that everything that we discover in a particular period of our history eventually goes to oblivion! I believe that there are strands of truth that carry over from one period of history to the next. These strands constitute our accumulated understanding in science. If this were not the case, if everything we understand today in science would dissolve tomorrow, then, as you say, science would not be any different than any other game, like a game of chess! It would not explain anything! However, I do believe that if we look very carefully at the evolution of ideas in science, ever since the earliest recorded investigations of the physical universe, we would see these persisting strands of truth.

MANNY: I really can't believe this, Jackie. But if you are insistent on trying to convince me, anyway, what are some of the examples of your 'strands of truth?'

JACKIE: An obvious one is the concept of the inertia of matter. In ancient Greece, Aristotle had the idea of the 'impetus' of matter. This is a sort of engine that keeps matter in motion, because of where it is in space. If, for example, a pebble is 'up' from the earth, its impetus would keep it moving 'down' to the earth's surface. About 16 centuries later, Newton discovered the concept of the inertia of matter, quantified by its so-called mass. This represented the resistance of the matter to a change of state of rest or constant motion in a straight line. This concept was built onto the concept discovered a generation earlier by Galileo, called the 'principle of inertia' — the idea that constant motion in a straight line is a natural quality of matter, not requiring any outside influence — contrary to Aristotle's idea. However, the amount of change caused in the matter's motion, for example its acceleration as caused by an external force, depends on the quantity of its inertial mass, which is an intrinsic quality of matter, according to Newton.

This concept of inertial mass, as a measure of a body's resistance to a change of its constant motion (or rest), is a direct carry-over from Aristotle's concept of the impetus of matter, as I see it, persisting for all of these centuries up to the present time.

When we discover new understanding in science, we can never be sure about which part of it will remain and which part will have to be discarded. But it is our obligation, as scientists, to keep on with the attempt to seek out new comprehension. Above all, the aim is not merely to play a game! The main aim is to add to our comprehension of the universe, however small this comprehension may be at any particular time in our history.

MO: I agree with this, Jackie. You may also add that this work of investigating scientific truths is not always great fun! There is a lot of hard, painstaking work! Most of it is up a blind alley — and this is painful! But some of it does pay off, little as it is. For this reason, I believe, we continue to get a thrill out of scientific research and teaching the subject.

MANNY: I agree with both of you, that scientific research is not always fun and that it is not a mere game! But I also do not believe, even when it is successful, that it goes into mysterious, underlying explanations. As I see it, science gives us useful correlations that were not previously known, that pave the way to even more correlations. Some of this leads the way to advances in technology, as we've seen since the earliest times in our history. I think that, perhaps, this is the ultimate aim of doing scientific work in the first place!

MO: Scientific research may help to advance technology, Manny, but this should not be considered as more than a by-product. If we evaluated the worth of science in terms of its contributions to technology, it would be a dead subject!

Don't you see, it would mean that scientific research would never be supported at the outset unless the final products in technology — a better car or washing machine, or more efficient products of war, etc. — would be possibilities!

As scientists, it is essential to make it clear to society, as well as to our students, that the only real aim of fundamental science is to increase our comprehension of the universe, in any of its domains, for the sake of knowledge alone!

JACKIE and MANNY: We agree with you on this in full, Mo.

On this agreement between the three physicists, it was time, once again, to break up, until next week. As they left the Conference Room, Manny, Mo and Jackie were feeling fortunate that they had this opportunity to think and to disagree with each other!

THE SEARCH FOR TRUTH IN SCIENCE: EINSTEIN VERSUS BOHR

MO: Well, here we are, colleagues, in the last week of the fall quarter. Do you think that we've made progress in our search for scientific truth? As for myself, I am more certain now on what the problems are that stand in the way of discovering truth in science than I was before we started our dialogues. I am still not certain, however, as to the correct approach to the truth. Perhaps there is no definite, correct approach!

JACKIE: I agree with you, Mo. After all, we are only finite beings, as we've remarked before, and the universe is infinite. That is to say, the range of its truths is unbounded. It then follows, as far as I can see, that we can never hope to achieve a complete understanding of any phenomenon of the universe, in any of its domains, whether in that of elementary matter or the problem of cosmology. If we can, at least, define the problems that stand in our way of increased comprehension, we would then be in a good position to increase our understanding, that is, to understand more today than we did yesterday.

MANNY: I have asked myself if my comprehension of the universe has increased since we started these dialogues, eight weeks ago. I believe that it has, though I feel frustrated in not yet succeeding in convincing the both of you that my views about fundamental uncertainty in Nature, based on the epistemological approach of empiricism and positivism, is the correct path to scientific truth. After all, this is the approach of the consensus in physics today! On the other hand, I am glad that I have not totally convinced you because I'm not entirely certain myself that your way, Jackie, of a totally ordered universe, based on the epistemological view of realism, is wrong.

I agree with you, Mo, that we should never abandon controversy, because, as we clearly see, it does lead to increased understanding.

JACKIE: I always knew that you were a very good scientist, Manny, and that you were therefore bound to feel this way.

Speaking of good scientists, Manny, who do you think is the most significant physicist of the 20th century?

MANNY: There is no doubt in my mind that it was Einstein — if you mean by his 'significance' one who gave us genuine progress in our comprehension of the physical universe, as well as pointing the way to new progress. He was not only the discoverer of the theory of relativity, but also one of the initiators of the quantum theory — these are the two essential cornerstones of modern physics.

JACKIE: If you think that Einstein is the most significant physicist of the 20th century, Manny, then why is it that you don't trust or believe any of the physics directions that he took during most of his professional career — the last two thirds of his life?[1]

MANNY: Because he was wrong! Everybody knows that Einstein was wrong about his physics directions in the later part of his life! He opposed the Copenhagen interpretation, initiated by Bohr and Heisenberg, because he was too old-fashioned in his thinking. Further, his concept of a unified field theory was obviously wrong! If it had been correct, he certainly had the brain power to prove it in his own lifetime. The unification that is being discussed today in elementary particle physics is along the correct direction — look at all of the brilliant people who adhere to it — while Einstein's idea of a unified field theory was followed by very few, though admittedly notable physicists as well as himself, such as Schrödinger, and Faraday in the last century. This proves that he was wrong, doesn't it Mo?

MO: Regarding the Copenhagen interpretation, I recall reading an interesting anecdote about a conversation between Einstein and Heisenberg. Heisenberg asked Einstein why he abandoned the operational view in physics, that influenced his original thinking in the theory of special relativity. Heisenberg said that it was Einstein who inspired his own views of the quantum theory in terms of the positivistic approach, and then, after it was formulated, Einstein changed his mind about the approach of positivism in science! Einstein replied to this question by advising that if one tells a funny joke once, it should not be repeated!

As to your attitude about Einstein, Manny, I really don't get it! On the one hand you say that he was the most significant physicist of the 20th century. But on the other hand, you say that he was wrong about most of his physics — especially the physics directions that he took in his mature stages! This seems to me to be quite schizoid, don't you think?

MANNY: Of course, Einstein did some very seminal work for 20th century physics — in his early work on relativity theory, the proof of the atomistic model of matter from his analysis of Brownian motion, his initiation of the old quantum theory in his studies of the photoelectric effect, his important studies of the probabilities of induced and spontaneous emission of radiation, and so on. But then, he went off base when the old quantum theory, that he was instrumental in discovering, matured to quantum mechanics and its interpretation given by Bohr and Heisenberg, and their collaborators, in the 1920s. Though theirs was a great discovery for science, Einstein refused to accept it! He stood still while the rest of the physics profession did accept it and moved forward!

JACKIE: Did Bohr's school prove conclusively that they had the correct path to scientific truth, Manny? I mean, in addition to the successes of the approach in atomic physics, were there any serious difficulties remaining to resolve?

MANNY: There were remaining technical difficulties — both physical and philosophical. But everyone agreed, as they still do today, that the ideas of quantum mechanics should be the correct path to scientific truth.

MO: Was Einstein's alternative approach to the quantum theory, in terms of a matter theory based on the theory of general relativity in all domains, including the microdomain, ever refuted?

JACKIE: The Einstein field approach to matter, in terms of general relativity, was never technically refuted, though it has yet to be fully exploited, to see just how true it may be to Nature. And Manny, since you admit that the Copenhagen approach has not yet been conclusively proven, in spite of its empirical successes, how do you know that it is not actually wrong in principle, regarding its conceptual approach? And why are you so sure that Einstein's theory is not correct?

MO: Perhaps what is true is that the correct approach to the objective truths of the universe is neither fully Bohr's nor Einstein's? Perhaps there are shreds of truth in both of their views!

JACKIE: That is quite possible, Mo. But if they are to be joined, it is essential to get rid of the parts of one of these theories that are logically incompatible with parts of the other theory that are retained.

MANNY: There we go again, Jackie, insisting on logical consistency in the laws of Nature! I still maintain that this is not a real necessity for science!

MO: We don't seem to meet on this point. I feel now, with Jackie, that the laws of Nature should be logically consistent in order to make any sense at all! Changing the subject, Jackie, could you tell me exactly what you meant by your term, 'significant scientist' in your question to Manny?

JACKIE: In my view, a significant scientist is one who asks significant questions, and then tries to answer them. The most difficult part of the charge, though, is to ask the significant question, and to know that it is indeed significant! That is, that its answer would give us increased comprehension of physical phenomena.

Of course, it is possible to ask insignificant questions in science and philosophy, and never be aware of just how insignificant they are! One may then spend the remainder of his or her life, as well as a large amount of the time of one's students and admirers, in trying to answer an insignificant question!

It is clear to me that both Bohr and Einstein were asking significant questions!

MANNY: How does one know, Jackie, when a question in science or philosophy is indeed significant? I mean, how do you know in advance, before starting to investigate it, that its answers would increase our comprehension of a physical phenomenon?

JACKIE: It is difficult to know in the initial stages of an investigation whether or not a question in science or philosophy is indeed significant. I think that, perhaps, one criterion would be whether or not the attempt to answer the question leads to more questions — if it does not, then most likely the question was not significant at the outset! I believe that this is one simple test of the significance of a question, especially in science and in the philosophy of science.

MO: I don't understand this, Jackie. If we believe that we have answered a question about some phenomenon in Nature, why should this answer necessarily lead to more questions?

JACKIE: There is an interesting anecdote about this, sometimes attributed to Bohr and Einstein. Niels asked Albert, "Why do you always answer a question with another question?" and Albert replied, "Why not?"

Seriously, though, the answer to your question, Mo, in my view, is that the amount of understanding that there is to be had about any phenomenon in Nature is unbounded, that is to say, it is infinite. On the other hand, we are finite beings. To have a complete understanding of any phenomenon in Nature would require us to have an infinite capacity of comprehension. Since this is impossible, since we can never become omniscient, we can never achieve a complete understanding of any phenomenon in Nature. Thus we are bound to have more questions about it after we have answered some of the questions, if they are significant. In this way, we continually proceed toward the objective truths of Nature, though we can never reach the limit of complete knowledge. It then follows that if a question leads to no more questions, it is most likely not related to the truth of natural phenomena, and therefore it was not a significant question in the first place!

MANNY: What you are saying, Jackie, is that if we think that we have no more questions to ask about a particular phenomenon in Nature, then in fact we do not really have any real comprehension at all about it! That is, we are wrong about what we think of as an explanation! I find this very hard to accept, Jackie!

JACKIE: Whether or not you can accept this, Manny, it is an important lesson of the history of science that we can never achieve a complete understanding of any phenomenon in Nature, just as Galileo asserted 300 years ago! This should not depress you, as a scientist! Our obligation, as scientists, is to increase our comprehension of the physical universe, little as it may be at any particular time, compared with all that there is to comprehend![2]

MO: One thing that you are saying, Jackie, is that, as scientists, we should be humble, not only for the purpose of placing ourselves in the proper perspective in the universe, but also for practical reasons. For if we should believe, at any particular time, that we had already learned all that there is to learn about any aspect of the universe, then we would no longer pursue any further comprehension of this subject. Thus, if we had not, in fact, completed our understanding of the subject, which you assert would always be the case when the subject involves Nature, then we would be halting our own future progress if we should stop the search!

JACKIE: This is correct, Mo. Such an attitude also implies that when the community of physicists should mistakenly believe that a subject is fully understood, a severe dogmatism about this subject would be instilled. That is, there would be a very strong pressure to stop the studies of any scientist or philosopher who may attempt to pursue a subject further, perhaps along directions not in agreement with the established 'truth' of it! It would not be unlike the reactions of the religious institutions of centuries ago to the studies of scientific inquiry, such as the Christian Church's opposition to Galileo and the other Copernicans of the 16th century. It is also very similar to the thought control of the Nazi and Communist regimes of this century, in halting free thinking in science and philosophy in the countries that they controlled.

MO: It seems to me, Jackie, that it is even worse today than it was under totalitarian regimes of our time, or the Church of the Inquisition period. Those oppositions to free thinking in science and philosophy came from the outside; it was external. However, the opposition to free thinking from the authorities of the fields of science and philosophy and their loyal followers is internal — it is anti-science within science itself! Unfortunately, it seems to be pervasive today, perhaps because of the size that the science profession has come to, and its great degree of organization.

MANNY: I really find this hard to take, friends! You are both claiming that in our own day — when the number of active physicists is much greater than the total number of all physicists who lived on Earth since the period of ancient Greece, when the number of printed pages in physics journals per month is thousands of times greater than it has ever been in the past, when the governments of the world, especially our own government, is spending many powers of ten more dollars on fundamental research than ever before — you are saying that these facts do not represent real progress in physics! This is truly ludicrous!

What you see, Jackie, as dogmatism in science, is really the necessary care that must be taken by our peers in order to protect us from the cranks! Don't you see, Mo and Jackie, we need this control if we are going to make real progress in our field of study! Otherwise, if anyone can think the way they please, we would have sheer anarchy, and there would be no progress at all!

JACKIE: There we go again. Manny, basing the truth or falsity of scientific results on consensus of opinion! This is clearly false! Look, Manny, if something in science or philosophy has not been demonstrated in a logical, scientific way, then no matter how many authorities and their followers believe in this as a truth, it is not any more true than if no one believed it! By the same token, if a truth is demonstrated by one person, in accordance with the rules of science and philosophy, then it would not be any less true even if all of the authorities on Earth and their followers refused to believe it!

Of course, the data coming from our measurements is true to these measurements! But exactly what these data are supposed to represent as an explanation for them is something else altogether!

MANNY: If I've said this once, I've said it a thousand times! Your so-called explanation is not more than an efficient mathematical and verbal way of representing the data at hand! We are not obligated, as scientists, to look for anything more, such as the underlying mysteries that you call 'metaphysics!'

JACKIE: You are wrong, Manny! You are even wrong about your own way of representing the data, especially in the fields of elementary particle physics and cosmology. The mathematical form that you use in your theory of elementary particles — that you mistakenly believe you can extend to cosmology — is very special. It is not general enough to include all possible explanations. Your mathematical language is in a form that can be interpreted as a probability calculus, easily applicable to a theory of measurement! It is a form that assumes at the outset that there are independent 'things,' apart from the apparatus that 'looks at them' — thus requiring a linear mathematical structure at the outset. The idea of wave-particle dualism is based on this particular mathematical form. These ideas are then, in part, your explanation of these data of elementary particle physics. What I am showing you is that you are not free of explanations — whether you like it or not!

Now, how do you know, Manny, that the mathematical forms that you are using to represent your data are not linear approximations for a totally different (nonlinear) field theory of matter, as in the case of Einstein's approach, based on general relativity, or in some other theory of matter, not compatible with the basis of the quantum theory? It is not that the different theories would predict the same data, so that one may choose one theory or the other, according to his or her whim! The theories are only making the same predictions when particular approximations for one of them is in use. However, when it would not be possible to use these approximations, the theories may make different predictions! Thus they would be separately testable from the empirical stand. The theory that agrees with all of the data at hand would be the one that must be accepted, if the other theories do not make all of these predictions. In this case, one explanation would have won out over the others!

MO: I see your point, Jackie. Manny cannot really justify his claim that all there is to science is the data!

Thus, Manny, I see that there are implicit explanations for the data in a way that they exclude other possible explanations — whether you agree with this or not! With this conclusion, it is the job of the scientist and the philosopher to determine which is the superior explanation — that is, the one that is more valid as a truth of the real world.

MANNY: What does this mean, Mo, "more valid as a truth of the real world?"

MO: Clearly, Manny, if one explanation has more empirical backing than another, and if it is logically and mathematically consistent, then it is a superior explanation. This has nothing to do with any popularity contest!

JACKIE: I think that this winds up our dialogues this quarter, gentlemen. Manny, you continue to take the positivist/empiricist stand, but I think that you are more aware of the meaning intended by the realist stand, that I espouse. Also, I think that I appreciate your view of science, even though I do not agree that it is sufficient! Perhaps I am too much of an optimist, as to what it is possible for us to comprehend about the universe, in any of its domains, in the final analysis.

I believe that you, Mo, still take a middle ground stand, epistemologically, relying on the evolution of the history of science to decide the truth for us. I see that you are trying hard to be fully neutral, unprejudiced and objective about your approach to the truths of the world. I cannot do this. Manny and I have definite intuitive feelings about the way that science should proceed. Perhaps our profession is best off to have the balance of the three of us, along with our mutual agreement to disagree!

The three physicists, Manny, Mo and Jackie, gazed out of the western window of the Conference Room for the last time, until the next quarter. It was one of those rare fall days when it was not raining in this part of Oregon. The particularity of the red brick buildings of the campus was etched out against the withering grass and chestnuts lying under the massive defoliated trees that bore them, and beyond this to the ever-rushing Willamette River — all seemingly separate things of Nature.

They turned briefly to gaze out of the eastern window, at the continuity of the golden fields, softly undulating in the gentle wind of that day. Then to the grandeur of the Cascades — those beautiful, gigantic volcanic structures of Nature looking down on Manny, Mo and Jackie, perhaps sympathizing with their struggle to know something of Nature that was already well understood by these less primitive manifestations of all that there is!

Definitions

[a] *truth* (p. 5)

It is interesting to compare the conflict between probability and order, as underlying the laws of Nature, with the two logical views of truth in Kant — 'analytic truth' and 'synthetic truth.' 'Analytic truth' is a necessary truth since it is a conclusion based on our invented axioms and logic, such as the truth of $2 + 3 = 5$. A 'synthetic truth,' on the other hand, is contingent, such as a 'scientific truth.' It is contingent because the axioms that one starts with are those of Nature — they are discovered by the scientist, rather than being invented. Since these axioms are tied to Nature rather than human invention, we can easily be wrong about their correctness. Indeed, it is the program of science to continually test the correctness of the axioms of the laws of Nature, by empirical verification and refutation, as well as by tests of logical and mathematical consistency. Thus, the laws of Nature, which are the goal of science, are synthetic (contingent) truths.

It is important, then, that these are two distinct types of 'truth.' But the distinction that Manny refers to, between the laws of probability and laws of order, is not of this sort. The question that Manny, Mo and Jackie discuss here concerns whether or not 'probability laws,' *per se*, can be 'scientific laws,' *per se* — i.e. whether they can be intrinsic in Nature, rather than analytic rules for interpreting measurements.

ᵇ *potentiality* (p. 15)

A physical possibility that does not become actual until it is 'actualized.' For example, a piece of chalk on the top of a table has a 'potentiality' called 'potential energy.' It refers to the possibility of converting its energy, by virtue of where the chalk is in space, on the table top in this case, into work. If one should slide the piece of chalk off the table top, it would fall to the floor and then do the work of breaking into pieces — its initial potential energy would then have actualized in the form of doing the work to break the atomic bonds that hold it together, thereby falling apart.

ᶜ *mu meson* (p. 26)

One of the elementary particles of matter, discovered in the late 1930s, as a component of cosmic rays. This elementary particle is very similar to the 'electron,' except that it is about 206 times more massive and it is unstable — it decays to an electron and two other particles, called 'neutrinos,' on the average in about a millionth of a second. It was discovered in the late 1940s that the mu meson is a decay product of a pi meson — a particle that plays the role of a 'glue' that binds neutrons and protons in nuclear matter.

ᵈ *time measure* (p. 28)

This is the parameter in physics that relates to a measure of duration. In classical physics, the time measure is an absolute (i.e. objective) feature of a material system, such as an ordering of the spatial trajectory of a particle of matter. That is to say, the change along the trajectory from one spatial point to another corresponds with a change of the time parameter. The time measure in classical physics is absolute in the sense that it is the same from any particular reference frame of observation.

In relativity physics, on the other hand, the 'time measure' is a subjective feature in the way of expressing the laws of the material system. That is, its value depends on the reference frame from which it is measured. In relativity, then, the time measure is not more than a language element that is used in order to express laws of Nature objectively, but it is not in itself an objective feature of the physical system.

^e *realist view* (p. 40)

This is the philosophy that asserts that there is a real world, independent of whether or not there are (human or any other type of) observers of it.

^f *holism* (p. 52)

The philosophic view of a truly closed system, that is, a system that is without separable parts.

^g *non-periodic* (p. 56)

This is in the sense that each 'year' of an orbit — the time it takes for the planet to return to an equivalent place in the sky relative to the sun's position — is longer than the previous 'year.' An example of such non-periodicity was first seen in the middle of the 19th century in regard to Mercury's orbit. Focusing on the perihelion point of the orbit (the place of the orbit that is closest to the sun) Mercury was found to increase the time taken each cycle to reach this point. This is known as the 'perihelion precession' of Mercury's orbit. It is equivalent to a slow rotation of the elliptical axes of the orbit.

^h *consistent theory* (p. 72)

When a theory in science is 'consistent,' it must make a unique prediction (qualitatively and quantitatively) for a given physical situation. For example, if a theory of gravitation should predict that the orbit of a planet about the sun's position is elliptical (an empirically true fact), but at the same time, without altering the physical conditions, the same theory predicts that the orbit is circular (an empirically false fact), then this theory is unacceptable as a scientific truth, because it is logically inconsistent, even though one of its predictions is true to Nature.

ⁱ *the principle of covariance* (p. 72)

This principle asserts that any law of Nature must be independent of any frame of reference in which this law is expressed, from the view of any other reference frame. If the reference frames are defined in terms of spatial and temporal measures, this principle implies that the 'transformations' of the space and time coordinates from one reference frame to any other, $(x, y, z, t) \rightarrow (x', y', z', t')$, must leave the forms of the laws of Nature, expressed with these respective space and time coordinates, unchanged — i.e. the laws expressed in these respective space-time languages must be objective.

^{j} *soliton* (p. 82)

This is a speculated elementary particle whose wave nature has the peculiar property that when it scatters from any object or other elementary particle it does not disperse into more waves. Such a wave was shown to follow from a particular sort of nonlinear wave equation. A nonlinear wave equation is one in which the wave solution appears in higher powers than unity. The idea was first considered in the 19th century when it was observed that shallow water waves may scatter from obstacles without dispersion. The nonlinear equation that predicts such waves was then formulated.

Ordinary waves, such as light waves, or the probability waves determined by quantum mechanics, are linear — the solutions appear in the equation only to the power of unity. In the latter case, the principle of linear superposition applies, implying that the arithmetic sum of any number of solutions of such an equation is another possible solution. This linear feature is essential for the structure of quantum mechanics to be interpreted as a probability theory. When such a wave scatters from any obstacle, it disperses into many new waves.

Notes

Foreword

1. A. Einstein and J. Grommer, "Allgemeine Relativitätstheorie und Bewegunggesetz" ["General Relativity Theory and Laws of Motion"], *Preussische Akademie der Wissenshaften, Phys.-Math. Klasse, Sitzungsberichte*, 1927, pp. 2-13.
2. L. de Broglie, *The Current Interpretation of Wave Mechanics: A Critical Study* (Elsevier, 1964), p. 43.

Preface

1. His views are clearly expressed in Plato: *The Republic*, Book vii. For a translation see W. H. D. Rouse (Mentor, 1956), p. 312.
2. Galileo, *Dialogue Concerning the Two Chief World Systems*, S. Drake, transl. (University of California, 1967), Second revised edition; *Two New Sciences*, S. Drake, transl. (University of Wisconsin, 1974).
3. J. M. Jauch, *Are Quanta Real?* (Indiana University, 1989).

Week 1

1. The concept of 'wave-particle dualism' is a pillar of the contemporary quantum theory. For a clear discussion see: D. Murdoch, *Niels Bohr's Philosophy of Physics* (Cambridge University, 1989); L. de Broglie, *ibid.*, Chapter 3; M. Sachs, "On Wave-Particle Dualism," *Annales de la Fondation Louis de Broglie* **1**, 129 (1976); A. Einstein, E. Schrödinger, M. Planck and H. A. Lorentz, *Letters on Wave Mechanics*, K. Przibram, editor, M. J. Klein, transl. (Philosophical Library, 1967), see especially, the correspondence between Einstein and Schrödinger.

2. A clear statement of the philosophy of quantum mechanics is in: B. d'Espagnat, *Conceptual Foundations of Quantum Mechanics* (Benjamin, 1971).

3. The predominant contemporary view is that the neutrons and protons — the constituents of nuclei — as well as the unstable elementary particles, such as the pions, are made up of more elementary units, called 'quarks.' The feature of quarks that Jackie refers to is their adherence to the so-called 'confinement theory.' This is the assertion that the further apart the quarks are from each other the stronger is their coupling; thus when they are almost in contact, they are 'asymptotically free.' This feature of their mutual force (opposite to that of nuclear, electromagnetic or gravitational forces) is to account for the fact that there is no direct empirical evidence for their existence as free quarks.

4. I heard this bit of satire on modern particle physics in the comedy routine of George Carlin.

5. Experiments that have investigated 'one photon at a time' are by: R. L. Pfleegor and L. Mandel, *Phys. Rev.* **159**, 1084 (1967). An interesting response to this experiment was by: L. de Broglie and J. A. Silva, *Phys. Rev.* **172**, 1284 (1968).

6. Plato, *The Republic, ibid.*

7. The realist philosophy of Einstein and Schrödinger has been discussed by them in the literature. See, for example, the correspondence between Einstein and Schrödinger in: Przibram, *ibid.* Also see: A. Einstein, "Autobiographical Notes" in *Albert Einstein — Philosopher-Scientist*, P. A. Schilpp, editor

(Open Court, 1949); W. J. Moore, *Schrödinger — Life and Thought* (Cambridge University, 1989).

Week 2

1. *Positivism* is a philosophical approach to knowledge in which it is claimed that the only meaningful reality is that which may be verified with the human senses.

   The initiation of this philosophy in quantum mechanics was Heisenberg's statement in his seminal paper, W. Heisenberg: "Über quanten Theoretische Umdeutung Kinematischer Und Mechanischer Beziehungen," *Zeits. für Physik* **33**, 879 (1925). An English translation is "Quantum Theoretical Re-Interpretation of Kinematic and Mechanical Relations," van der Waerden, editor, *Sources of Quantum Mechanics* (Dover, 1968). Heisenberg opens the article with the following statement:

   > The present paper seeks to establish a basis for theoretical quantum mechanics founded exclusively upon relationships between quantities which in principle are observable.

2. A *single-valued logic* is one in which a statement is either true or it is false, but not both simultaneously. A *many-valued logic*, such as a *three-valued logic*, [H. Reichenbach, *Philosophic Foundations of Quantum Mechanics* (University of California, 1944)] proposes a system of three possibilities as truth values — true, false or something in between. A rational generalization of this logical system would be based on an indefinite number of truth values for any given statement. Whether or not this sort of generalized logical system would be useful for the laws of Nature has yet to be determined.

Week 3

1. Galileo's discovery of the relativity of all motion is revealed in his statement, in *Dialogue Concerning Two Chief World Systems* (Preface, Ref. 2, p.114), where he says:

There is one motion which is most general and supreme over all, and it is that by which the sun, moon and all other planets and fixed stars — in a word, the whole universe, the earth alone excepted — appear to be moved as a unit from east to west in the space of twenty-four hours. *This, in so far as first appearances are concerned, may just as logically belong to the earth alone as to the rest of the universe, since the same appearances would prevail as much in one situation as the other.* (My emphasis).

This statement clearly reveals Galileo's belief in the relativity (subjectivity) of *all* motion. In his view of motion, then in a specific example, it would be equally correct to say, as Aristotle did, that the sun moves around the earth, as it would to say, as Copernicus did, that the earth moves around the sun. Thus, Galileo went beyond the Copernican heliocentric idea that all heavenly objects move around a fixed sun, because in Galileo's view, there cannot be *any* absolute (fixed) point in space.

2. There is a voluminous literature on the problem of the twin (clock) paradox. I have discussed it in: M. Sachs, "A Resolution of the Clock Paradox." *Physics Today* **24**, 23 (1971). A response to my article and my reaction to it is in the *Letters Column, Physics Today* **25**, 9 (1972).

Week 4

1. In a personal letter that Einstein wrote to de Broglie in 1954, (L. de Broglie, *Annales de la Fondation Louis de Broglie* **4**, 56 (1979)), he said:

    Die gravitationsgleichung waren *nur* affindbar auf Grund eines rein formalen Prinzips (allgemeine Kovarianz), d.h. auf Grund des Vetrauens auf die denkbar grösste logische Einfachheit der Naturgestze. [The equations of gravitation were able to be discovered *only* on the basis of the conviction that the laws of Nature have the greatest imaginable logical simplicity].

2. In a lecture that Paul Dirac gave, during the 1979 Einstein centenary year (G. Holton and Y. Elkana, editors, *Albert Einstein: Historical and Cultural Perspectives* (Princeton University, 1982, p. 351), he said:

    It seems clear that the present quantum mechanics is not in its final form. Some further changes will be needed, just about as drastic as the changes made in passing from Bohr's orbit theory to quantum mechanics. Some day a new quantum mechanics, a relativistic one, will be discovered, in which we will not have these infinities occurring at all. It might very well be that the new quantum mechanics will have determinism in the way that Einstein wanted. The determinism will be introduced only at the expense of abandoning other preconceptions that physicists now hold. So, under these conditions I think it is very likely, or at any rate quite possible, that in the long run Einstein will turn out to be correct.

3. I have discussed their philosophical differences in: M. Sachs, *Einstein Versus Bohr: The Continuing Controversies in Physics* (Open Court, 1988).
4. Kepler's metaphysical views are discussed in: A. Koestler, *The Sleepwalkers* (Penguin Books, 1959), Part 4, Chapter 2. Faraday's metaphysical views are discussed in: J. Agassi, *Faraday as a Natural Philosopher* (University of Chicago, 1971).

Week 5

1. The concept of holism in the quantum theory is discussed, for example, in the articles by H. Stapp and D. Finkelstein, in: R. F. Kitchener, editor, *The World View of Contemporary Physics* (State University of New York Press, 1988). A critique of their views is in: M. Sachs, *Philosophy of Social Science* **20**, 233 (1990).
2. A discussion of holism in general relativity is given in: M. Sachs, *ibid.*

Week 6

1. C. P. Snow, *Two Cultures and the Scientific Revolution* (Cambridge University, 1964), Second edition.
2. For a discussion of Dirac's view of beauty in science, see: H. S. Kragh, *Dirac: A Scientific Biography* (Cambridge University, 1990), Chapter 14.

3. On Einstein's concept of simplicity in science, see his "Autobiographical Notes," in P. A. Schilpp, editor, *Albert Einstein — Philosopher-Scientist* (Open Court, 1949), p. 69.
4. E. Hubble, "The Exploration of Space," in M. K. Munitz, editor, *Theories of the Universe* (The Free Press, 1957).
5. I have discussed this feature of the theory of general relativity, in a mathematical fashion, in: M. Sachs, *General Relativity and Matter* (Reidel, 1982), Chapter 7.
6. There is further qualitative discussion of this view in: S. Hawking, *A Brief History of Time* (Bantam, 1988).
7. J. Silk, *The Big Bang* (Freeman, 1989), revised, updated version. Also see Hawking, *ibid.* A more technical discussion of the inflationary model is in: A. H. Guth, *Phys. Rev. D* **3**, 347 (1981).

Week 7

1. See Note 1, Preface.
2. M. Maimonides, *The Guide of the Perplexed* (University of Chicago, 1963), S. Pines, transl.
3. Schrödinger's criticism is discussed in: M. Sachs, *Einstein Versus Bohr* (Open Court, 1988).
4. On this question, see: T. S. Kuhn, *The Structure of Scientific Revolutions* (University of Chicago, 1970), Second edition.

Week 8

1. I have discussed this question in: M. Sachs, "Einstein and the Evolution of Twentieth Century Physics," *Annales de la Fondation Louis de Broglie* **16**, 241 (1991).
2. In this regard, it is interesting to quote from a prayer that was written in the Middle Ages, by Maimonides, for the scientists of his day:

Grant me strength, time and opportunity to correct what I have acquired, always to extend its domain; for knowledge is immense and the spirit of man can extend infinitely to enrich itself daily with new requirements. Today he can discover his errors of yesterday and tomorrow he may obtain a new light on what he thinks himself sure of today.

In the Renaissance period, the Swiss philosopher, Jean Jacques Rousseau, said:

Le monde réel a ses bornes, le monde imaginaire est infini. [The world of reality has its limits, the world of imagination is boundless].

This comment indicates that our imagination is bound to go off on many unreal tangents in our pursuit of the truths of the universe. It is our obligation, then, as scientists and philosophers, to continually test paths of inquiry, being ready to reject them if they do not meet the standards of empirical verification and logical consistency, as in Maimonides' prayer. In our own day, the importance of the refutability of a scientific theory for progress of our knowledge has been emphasized by the philosopher, K. R. Popper, as discussed in his: *Conjectures and Refutations* (Basic Books, 1962).

Epilogue

AN ESSAY ON DIALOGUE

Robert R. Sachs
Bion Environmental Technologies, Inc.
Amherst, New York

For those with serious interest in the way of Dialogue, it is necessary at a certain point in time to reflect on some of the fundamental principles of Dialogue itself. We might say initially that for us as human beings Dialogue is an all-pervasive feature of our existence, given that we conceive it to include in addition to verbal language, non-verbal communication and thought, which we might call internal Dialogue. As such, the only experience that may possibly lie outside the bounds of Dialogue is dreamless sleep, which supposedly occurs without any thought.

As for the process of Dialogue, we must note that each particular experience (or example) of Dialogue is unique, fundamentally different from all others. This is due to the fact that the individuals engaging in Dialogue are always changing, whether it be different individuals or the same individuals at different times. As such, Dialogue can never be completely systematized or exhaustively analyzed. It always involves a creative element that defines our attempts to predict exactly where it will go and what conclusions it will reach. Each Dialogue is an adventure that takes its interlocutors to places they can never fully anticipate. In short, the unpredictable is an inextricable feature of Dialogue.

All experiences of Dialogue do, however, have some elements of commonality. They all require the presence of (at least) two distinct selves and an interaction between (or among) those selves. Dialogue therefore includes an element of discreteness as well as one of continuity, neither of which can be reduced to the other, the former in terms of the selves and the latter in terms of interaction. This leaves us with an apparently impossible situation involving the contradiction of continuity versus discreteness. Yet while our reason may have a problem accepting this state on contradiction, it can't invalidate Dialogue itself, which somehow embraces it. As such, Dialogue transcends normal reasoning. It does not give us a new way of thinking but rather a new way to approach thought. It makes thought limit itself by making clear to it the contradiction that is eternally present at the horizon of the approach taken.

In view of the contradiction present in Dialogue, we are further constrained in our attempt to conceptualize it. At best we can point out some of the basic aspects of Dialogue, coming to terms with its rich but contradictory nature. These aspects come to light through a contemplation of the nature of self and of interaction in Dialogue. The first aspect focuses on the role of the self and is represented by the principle of power; the second aspect deals primarily with the nature of interaction, represented by the principle of love; and the third aspect focuses on the way the self and interaction are grounded within Dialogue — this aspect is represented by the principle of freedom.

All three aspects, identified by power, love, and freedom, are in truth interdependent, i.e. not really separable except for the sake of analysis. Furthermore, the aspects of Dialogue operate in all Dialogue, from mundane conversations to the most abstract metaphysical contemplation. This is due to the fact that while the content of Dialogue may change, the fundamental workings of Dialogue remain the same, involving selves and the interaction between them.

A. Dialogue Under the Aspect of Power

All Dialogue involves a dynamic tension between (or among) the selves engaged in it. Without this tension, there could be no Dialogue because the opposition of distinct selves provides the raw material that keeps the flame of Dialogue going. Were the selves to be totally consumed in the interaction, Dialogue would cease to exist, thereby replaced by a state of homogeneous unity.

As such, though the selves engaging in Dialogue must give up something of themselves to participate in Dialogue, they cannot exhaust themselves entirely without destroying that to which they contribute. This means that there is always a necessary holding back by the self, thus preserving the identity of the self. The self then, in order to maintain its identity, must have a hidden side, i.e. something that remains outside of the interaction of Dialogue.

Since the hidden side of the self is always in relation to its manifest side, the nature of the self is most accurately characterized by dual and codependent processes of holding back and letting go, and is a kind of balance struck between the two. This balance can also be described as a flexibility in moving between the states of manifestation and hiddenness. The balance or flexibility of the self, I, will identify as the power of the self.

Every self must have some power in order to exist as a self, but not all selves have the same degree of power. We measure power by the amount of flexibility found in the self, which is the self's ability to remain the same while undergoing various changes, the most basic of which are those involving shifts to and from the states of manifestation and hiddenness. A very powerful self remains unmoved, i.e. in a state of balance, even when experiencing very strong forces, as in a severe personal crisis like a confrontation with death. A weak self, on the other hand, is constantly tossed about by the winds of change, having extreme difficulty finding a place of balance. Most selves rest in between these two extremes.

Let us next consider some of the basic kinds of forces that create imbalance within the self, for the acquisition of power results from the process of mastering these forces.[1] These forces produce various movements to us as (human) activities. The process of mastering the forces is therefore equivalent to mastering these activities. Though it is not really possible to single out every force (or activity), we can group them roughly in terms of four basic categories: the theoretic, creative, praxitic and reflective forces. Each of the four forces implants within the self a particular kind of desire, which the self then proceeds to actualize through some sort of activity.

The theoretic force produces within the self a desire to seek order and lawfulness, to pin down the flux of change with schemes that identify fixed structures and regularly recurring patterns. This becomes manifest to us as the activities of science and the other scholarly pursuits. The creative force implants within the self a desire to undo that which is and create something new. In contradistinction to the theoretic force that posits static forms, here change is completely embraced because the creative act requires the ever-flowing presence of change. We come to experience the creative force through activities of a creative nature, usually in the form of the creation of music, poetry, and other works of art, but also in the creation of scientific and philosophic theories.[2]

The praxitic force gives rise to a desire to engage in praxis, i.e. an application of the formal rules to a practical situation. Through the praxitic force, the self (usually) seeks out other selves in order to work on certain projects of an essentially communal nature. Here the self fulfills itself by participating in group activity, as a member of a team whether it be sports, construction, or a medical research team.

Lastly, the reflective force engenders within a self a desire to explore the fundamental principles and presuppositions of whatever is present to the self, including everything from philosophic schemes and scientific theories to personal value systems. In all cases the reflective force urges the self to locate the natural limits of the object of focus, and thus go beyond it. This consequently enables the self to have the choice to remain with the object of interest or move onto something else. The activity manifest through the reflective force is philosophy, perhaps best exemplified in the activities of Socrates.

The forces and correlative activities show the various directions in which the self can move to acquire power. The usual acquisition of power does not, however, automatically follow upon one's decision to take a particular course of action. Power is never freely given; it must be taken by force of will and conquered. In the process of acquiring power, the self is continually confronted by opposition which it must face and overcome. Such opposition may be a theoretical problem to be solved, a block of marble to be sculpted, a moral dilemma to be resolved, or a wrestling opponent to be fought and conquered. In all cases, the struggle with some sort of opponent is inexorable; and as soon as a particular opponent is defeated or the battle concluded as a standoff, another will immediately take his (or its) place. This process goes on as long as the self exists.

Given the fact that power can only be acquired through confrontation with an opponent, we must next determine exactly how such acquisition occurs.

The best approach to this problem comes through the analysis of the simplest and most direct form of confrontation — physical combat. The three basic skills of a successful fighter are the attack, parry, and the feint. The attack, when it is not neutralized by the enemy's defenses, is indicative of a successful penetration of the opponent's sphere of influence. The parry signifies one's ability to maintain one's sphere of influence in the face of attack. And the feint is a means of making the opponent expose new areas for attack within himself.

In the broadest sense these are the necessary skills in all exercises of power. Through the attack one penetrates and this grasps the essence of the opponent's ways. In the parry, one reaffirms the validity of one's position (and in particular in terms of one's basic premises and presuppositions) through a successful defense of it. Lastly, in the feint, one uncovers new and yet unknown parts of the opponent which one can then begin to probe.

Finally, we must deal with the purpose of power. This is especially important because the purpose is frequently forgotten when the self becomes enthralled by defeating opponents, and usually other people. In this case, the self loses sight of the significance of opposition, and consequently the true nature of power itself. As we originally defined power, it is the state of maintaining balance while undergoing change, or put another way, it is the flexibility of the self acquired through the mastery of the forces that affect the self. As such the true purpose of power is not the mere conquest of opponents, which only leads to arrogance and an unquenchable thirst for more and greater victories; but rather the process of acquiring greater balance and flexibility accompanied by more and more mastery of that which affects the self. Opposition to the self arises for the ultimate purpose of helping the self achieve this state in the fullest way possible. Any lesser goal will never embody true power, and consequently never bring true fulfillment.

In the domain of twentieth century physics, we find a situation in which the power of the self can be vigorously tested. This is due to the fact that in this century, unlike previous ones, there is a multiplicity of contemporaneous theories, each of which finds a number of advocates to sustain it. As a result there are incommensurable divergences among the theories, which are in turn experienced by the physicists drawn to one theory or another.

In times prior to this century, a single physical theory was dominant for a particular span of time, Aristotelian physics in the Middle Ages, and then at the later period of the 17th through the 19th centuries, Newtonian physics. During these times a self would encounter opposition, but because of the unifying force of a single overarching theory, opposition could be ultimately discounted. All differences would be brought together under the all-encompassing umbrella of a particular approach to the physical world.

Today there is no single umbrella in physics. There are instead a number of theories, each claiming to be the true representation of the workings of Nature. The two main theories are the relativity and quantum theories. Due to fundamental differences between these two theories, including the basic philosophic ones of continuity versus discreteness and realism versus operationalism, the two are completely dichotomous, presenting us with a state of true opposition that will never be resolved.

Though it might not appear so at first sight, given this state of irrevocable opposition, the selves committed to either relativity or quantum theory can be greatly aided by the presence of the opposing theory represented by its advocates. The reason is that the opposition forces one to deepen one's understanding of one's own approach, viz. how much it can and can't do. This is very clearly brought to light when two selves from opposing camps engage in Dialogue. Here each interlocutor used the three interrelated skills of combat — the attack, parry, and feint — either to point to the limitations of the opponent while manifesting one's strength; to defend one's approach against the attacks of the opponent; or to uncover new areas of difference which one can then attack. If all is done for the sake of true power, then such conflict is not at all just a pleasant diversion from the true work of science, but rather a valuable component in the ongoing advancement of the field.

B. Dialogue Under the Aspect of Love

All Dialogue involves interaction (or communication) between the selves participating in it. Interaction in turn implies the existence of some common ground, without which the selves encountering one another would be like ships passing in the night, experiencing no exchange. The common ground is that which binds together selves, and is best characterized by the principle of love. Unlike power in which opposition is taken to be most real, in love opposition is considered illusory, indicative of a partial, incomplete view of things. The true picture is one in which opposition is assimilated within a comprehensive whole that contains no separable parts. The assimilation of opposition will, of course, only find its full realization in the ultimate unity of all things. Though this cannot ever really be experienced, it is nevertheless an ideal that guides us in our various attempts to grasp it. Without this ideal there is no assurance of the ultimate assimilation of opposition, meaning that opposition is real and unity (or love) only an illusion. The presence of ultimate (or infinite) love therefore grounds and validates all levels of finite love, i.e. the unity which is experienced by finite beings, viz. selves.

We should note that within love selves retain their integrity as distinct individuals but are no longer separate. They are rather like modes of a continuum, all partaking of one another and of the whole of which they are a part. In love, selves interact in a cooperative, non-antagonistic way in the pursuit of common goals and a single truth, all in an effort to imitate the state of ultimate unity. Yet (as we have said) since it is not possible to experience this state directly, we can at best reach a state of imperfect unity, which is always set over against some opposing system. This is the inevitable condition of the finite in that no matter how much assimilation has taken place, opposition is still present. But given the ideal or ultimate unity, such opposition can always be interpreted as a sign of the incomplete nature of finite things, i.e. illusory in terms of infinite love.

In sum, guided by the principle of love the self is able to experience something greater than himself by being part of a larger whole. By the same token, he also recognizes his own limitations in that while the whole flows through him, as it were, his experience of the whole is only from a particular, limited perspective. No finite self can experience or know the whole from all perspectives. The finite self thus encounters opposition in the form of other perspectives. Yet since all perspectives look upon the same continuum of Being, there never really is any true and unresolvable opposition between or among them, given that they are all modes of the same unitary whole. In short, while reality is one, it may be seen in many different ways.

One of the most important and interesting finite forms of love is (human) language. Language is a comprehensive unity with many instantiations (i.e. parts), including the various languages of the world as well as individuals' perspectives on a particular language. Though there may be differences between the different instantiations of language, as for example between English and Japanese, given the ultimate unity of language, it is always possible to translate from one instantiation to another. This in turn necessarily presupposes some common ground which incorporates the opposing perspectives and makes translation possible. The absence of such common ground would put all instantiations of language in jeopardy for it would mark the end of translation and thus of communication itself.

In short, since all human beings, no matter how diverse their perspectives, can come together to communicate through language, there must be an ultimate common ground which facilitates all possible juxtapositions of instantiations of language. Though this ultimate common ground can never be fully realized (or experienced), it is nevertheless a necessary precondition of all uses of language.

We should also note that since language itself is a finite form of love, it is part of a larger, more comprehensive whole, which we might characterize as Being. Within the circle of Being language becomes a particular perspective (or mode) set over against that of reality. As parts of the same ultimate unity, language and reality reflect the same order of things but in two different ways. The differences in perspective can be overcome through the power of translation made possible through the common ground of being. (These ideas are akin to the Spinozistic conceptions of the modes of mind and body).

The principle of love is the underlying ground of all human activities and endeavors because it holds together all the particular actions, thoughts, and words that make up those activities and endeavors. Love is the source of their objectivity. Without love they would disintegrate like a body stripped of its soul at the point of death, giving rise to a state of relativism with each person going his own way, and ultimately to the state of solipsism. Love also binds together all the different activities and endeavors by acting as the common ground in which they all become parts. This is an ideal unity but it nevertheless impacts each and every one of our activities and endeavors, if only subliminally.

A specific example of an activity embodying the principle of love is the science of physics, identified as the study of the fundamental laws of the physical world, or Nature. As in all sciences the goal of physics is complete knowledge of its subject matter. This ideal limit guides the physicist in his various activities and becomes known to him through the scientific advancements that he achieves. In other words, though the state of complete knowledge is only an asymptotic limit, it nonetheless is experienced (at least in part) in the step by step progress made toward the realization of the ultimate end. In short, the principle of love, here representing the entirety of knowledge of the physical world, must be present at every stage of the activity of physics, to furnish the necessary unity without which it could not exist.

Given the guiding force of love, progress in physics is assured, though not subject to any regular schedule. As for the measurement of progress, this takes the form not of a straight continuous line but of a jagged one with stops and starts. The reason for this is that at various stages in the history of the science, there have been major shifts in orientation, represented by the emergence of a new physical theory, and accompanied by the discardance of a long-established one that has not proven itself effective in handling new problems. We find this in the emergence of the Galilean and Newtonian theories in the late 16th and early 17th centuries which dethroned Aristotelian physics, and then in this century with the emergence of Einstein's relativity theory and Bohr's theory of quantum mechanics at which time the classical approach of Newton lost its place of pre-eminence.

It might be thought that the shift to a new physical theory is a truly revolutionary change, marking a total break with the past. But such changes need not be interpreted just in this way. There are essential and often incommensurable differences among various physical theories, including such things as the nature of matter and force, and the significance of measurement. It is possible, nevertheless, to find threads of commonality that span the differences. For one thing, all physical theories attempt to explain physical phenomena, even though the phenomena may be interpreted in highly divergent manners. This is coupled with the fact that all physical theories must be verified through experiments that test the predictions they make. Secondly they must be organized in a logically consistent and coherent manner.

In terms of the principle of love, the truth actually lies somewhere between the two poles of commonality and difference. This is due to the fact that while each (proven) theory has a universal character measured by its considerable power of prediction, each is also a finite perspective that grapples with the infinite by creating a cross-section of it. Thus, in a sense, the vast expanse of Nature is comprehended, though never in its entirety since the power of explanation (which is embodied in particular theories) must begin with a finite set of axioms and then move in a particular direction opposed to others.

This opposition is not a hard and fast one because the different theories, under the aspect of love, are perspectives on a whole that contains them all. To use another analogy, they are like grooved tracks made in the earth, all of which are separated from one another but also unified by the same earth that underlies them.

Seen in this way, the history of physics presents us with various perspectives, usually with one taking a dominant position at a particular time (though in the 20th century there are two). But as perspectives, they must all be part of the same ultimate unity, and in temporal terms, part of the same quest for truth (even though such truth appears *to us* under different guises).

In the final analysis, though differences always arise among individuals committed to divergent perspectives, (in this case physical theories or versions of those theories), given the principle of love, there is always a way for individuals to get together, set up some common ground, and then work cooperatively toward the realization of some commonly held goal. This is not to say that differences ought to be abandoned. Differences among individuals are important and should be manifest but they must be seen in the proper light, i.e. in terms of perspectives on the same continuum of Being, or as modes of a single unified field. As such differences will serve the cooperative effort rather than create unnecessary opposition.

C. Dialogue Under the Aspect of Freedom

The third aspect of Dialogue involves movement to and from the ground of Dialogue. This ground is in the form of basic presuppositions and axioms underlying the positions taken. The ground is not normally manifest but we come to know it through an application of the presuppositions and axioms to a particular problem that needs to be resolved.

The ground is set up through a hierarchical ordering of values, with some things valued less, some more, some considered the most rudimentary values, while still others discounted as having little or no value. This ordering of values we can characterize as a value system. Value systems are all-pervasive in human existence, determining the nature of all our decisions and actions on a personal, cultural, and universal level. Value systems also underlie scientific theories and philosophic systems which structure reality according to a prescribed ordering of values, i.e. specific presuppositions and axioms as opposed to others. In short, value systems entail a kind of selection process that forge a particular ontology and patterns of thought best suited to that ontology.

As for the movement that occurs to and from the ground, it is in effect an act of either focusing or disengaging from the object of focus. In terms of the discussion of value, it is the act of turning one's attention toward or away from either the underlying values of an approach or the applications of the value scheme in the resolution of specific problems. There is in truth no clear line of demarcation separating the values from the application of the value system, though in the act of focusing we can only move toward or away from first principles. In other words, it is not possible to apply one's principles while at the same time reflect upon them.

The act of focusing/disengaging is unlike anything else that we do because it alone is done freely. All other actions (including thoughts) are inextricably bound to some value system that determines through a particular hierarchy of values the direction we will take in our thoughts, words, and deeds. Further, given this analysis, it follows that freedom has an ontological status, being an essential feature of human existence. For irrespective of the material conditions, all human beings have the power of freedom and constantly exercise it through the act of focusing/disengaging. This is what truly distinguishes the human being from the animals. The latter are determined to follow the particular value systems that order their lives and behavior. They are not free to focus their attention on those values nor disengage their attention from a strict application of the rules.

While all human beings experience freedom to some degree, insofar as it is an essential feature of humanness, very few come to know the full realization of freedom, even though the latter is in truth the necessary condition for all experiences of freedom. This full realization begins when one turns one's attention to the first principles of one's value system. Serious contemplation of these principles results in an awareness of their hypothetical nature, i.e. that they do not constitute the truly solid foundation one might have expected them to be. Their power actually lies in the facility with which they can be used in various applications. But the principles in themselves are mere hypotheses and thus of an ultimately arbitrary nature.

Such awareness propels one into the state of a very peculiar kind, one likened to the Absurd of the existentialists. This is a place in which all values come together, which is equivalent to a state of valuelessness since value is only meaningful in the context of a particular value system, with a specific hierarchy of values.

In the state of valuelessness such things as truth and reality are no longer relevant or meaningful. Reality is only meaningful when there is an underlying ground that remains fixed throughout change. But in the state of valuelessness there is no unchanging ground because all 'grounds' or basic value orderings co-exist relative to one another. None can be valued over any other. The same problem arises concerning truth, in that without the possibility of a set of facts to be confirmed or the existence of a coherent whole, truth is meaningless. And we find neither facts nor a coherent whole in the state of valuelessness.

The experience of valuelessness can be either a positive or a negative one. When it is taken negatively, valuelessness results in a vision of the utter pointlessness of all value systems in that they are all essentially equivalent with no way to choose one over another. This in turn is due to the fact that all criteria for determining 'value judgements' arise within particular and essentially discrete value systems. In other words, these criteria themselves are value dependent. One who experiences the negative side of valuelessness is immersed in a state of paralysis, unable to take any action because there are no longer any values to stimulate one to move in a particular direction.

When valuelessness is, on the other hand, interpreted positively, the individual experiencing it opens himself to complete and radical freedom that becomes in turn a principle that stirs one to activity. This state of freedom involves an awareness of the way valuelessness is ever-present on the horizon of value. What this means is that while we, as human beings, always take on some value orientation or other, (all) other values or value orientations are set over against the one we have chosen. These other values literally constitute an infinite ocean of possibility out of which a small number of finite value systems become manifest. In freedom we learn to work with the ocean of possibility to make it serve our ends, as opposed to fighting against its encroachment on our territory (which is not really possible in the first place).

In sum, radical (or true) freedom gives rise to the unique posture of detached attachment, in that while we must be continually involved in some value system, we are also in some way part of the boundless ocean of all-values-together, and thus detached from any particular value system.

The advantages of true freedom are twofold. In the first place, it enables us to move at will in and out of value systems, unlike the lesser forms of freedom in which individuals are perennially ensnared by a particular value system and thus unable to detach from it in order to explore other systems. Secondly, it gives us greater power to penetrate and grasp the essence for any value system. This is due to the fact that given the position of detached attachment, a value system is approached by means of its fundamental presuppositions, i.e. in terms of what it can and can't do. Once the limits of a value system are exposed, we have a better chance of seeing it for what it is and not reading into it things that do not really belong there. For individuals attached to a single value system, there is often no reflection on the limitations of the approach. As a result, attempts are made to apply the system in ways that will only end in failure.

Finally let us consider the way the principle of freedom applies to such activities as the study of physics. When considering the issues of physics either in personal meditations or through conversations with others, freedom enables one to expose the first principles of one's approach so that one is then in a position to detach from the approach (i.e. the value system) when it is no longer able to handle a particular problem under investigation. Without (true) freedom one is determined to remain with the same ground even if it can no longer resolve contemporary problems.

The great physicists are those who have been able to tap into this freedom. As a result they were able to detach from the pre-eminent theory of time, when that theory proved inadequate, and then by means of freedom made manifest a new theory, i.e. value system, out of the ocean of all-values-together.

It should not be thought, however, that true freedom is only possible for a privileged few. Since it is a feature of all humanity, more individuals could exercise it (and thus make significant changes in the field) were they willing to make the effort to return to the ground of their given value system, i.e. begin to reflect on first principles.

A possible way to stimulate more reflection of this sort is through Dialogue of a philosophical nature that attempts to uncover the ground of one's approach through the incessant (and sometimes annoying) process of questioning. Though it is often difficult, philosophical Dialogue will inevitably lead one to greater depths of freedom, manifest on the one hand as greater facility in penetrating and resolving problems and on the other, in bringing to light new ideas and ultimately new approaches within the field.

Notes

1. We should note that true power, i.e. complete mastery of the forces, is only an ideal limit since the extent of the forces and correlative activities is virtually infinite.

2. There is in truth an essential dovetailing of the forces. They are never totally distinct, all impacting one another to some extent. There are nevertheless identifiable lines of direction determined by the basic kinds of forces. Scientific activity, for example, can be creative, but it is still more science than art.